For Every Voice, A Different Truth

For Every Voice, A Different Truth: An Advisory Handbook by Tenth Grade Students at The Institute for Media and Writing at the Bayard Rustin Educational Complex

ISBN: 1-932948-34-1

Classroom Teacher: Erin Quigley
Supporting Classroom Teachers: Tracy Faber, Leila Sinclaire, Cynthia Moore

SPI Director: Erick Gordon
Curriculum Consultant, SPI: Kerry McKibbin
Site Assistant, SPI: Leigh Reilly
Production Team, SPI: Chris Fazio, Jim Fenner
Teaching Artist, SPI: Laura Starecheski
Cover Design: Hugo Ortega
Cover Layout: Chris Fazio, Jim Fenner
Editors: Kerry McKibbin, Erin Quigley, Leigh Reilly, Leila Sinclaire

For more information, please contact:
publishspi@gmail.com
Student Press Initiative/MCPET
Teachers College, Columbia University
Box 182
525 West 120th Street
New York, NY 10027

Contents

Acknowledgments

All of us who worked on *For Every Voice, A Different Truth* would like to extend our appreciation to:

John Angelet, John Loonam, Tracy Faber and all faculty and staff at The Institute for Media and Writing at the Bayard Rustin Educational Complex.

From Region 9: Peter Heaney, Superintendent, and Elaine Gorman, our Local Instructor Superintendent.

At Teachers College: Ruth Vinz, the Morse Professor of Teacher Education and Director of the Center for Professional Education of Teachers at Teachers College.

From the Institute for Student Achievement: Gerry House, President and CEO, Steve Stoll, Monica White and Liz Johnson.

Jackie Ancess, Co-Director, National Center for Restructuring Education, Schools, and Teaching (NCREST) at Teachers College, Columbia University.

And, of course, from the Student Press Initiative at Teachers College: Erick Gordon, Director, Jim Fenner and Chris Fazio, the Production Team. Without these three, this publication never would have happened.

Lastly, a tremendous thank you to all those incredible, inspirational people out there who gave of their time to be interviewed by our students and to share their experiences, perspectives and insights… you are the heart and soul of this book.

Foreword

One of the sillier things people say about school has to do with distinguishing between the classroom and the real world. You have probably heard people refer to graduating or leaving school as "going out into the real world." As if there was something less real about your life in school. Is leaving the house at seven in the morning any less real because you are carrying schoolbooks? Are the people that jostle you in the halls less real than they would be on the street? Are the decisions you make everyday about your future any less important? The questions you face, the skills you use, the challenges you overcome, are no less real because they take place in a building dominated by teenagers. The thinking you do and the ideas you express are no less important because you are working at deepening your thoughts and developing your expressive abilities. Successful adults, in every field, go through their days trying to deepen their thinking about the world and strengthen their skills at meeting its challenges. That is what the real world is like.

This belief is one of the things that made me want to be director of The Institute for Media and Writing in the first place. I want to be part of creating a school environment that recognizes students as people with important questions and important ideas. I want to help teachers create classrooms where students can do real work that is important to them now, in addition to helping them build skills that will be important to them later. I am proud to see this work from IMW students, just as I am proud to work with teachers like Erin Quigley and Tracy Faber. They know their students live in the real world.

One of the things I love about this book is that the authors know it too. They take their ideas and experiences seriously, and understand that thinking and writing about these ideas and experiences is real work, as important as any piece of work anyone of any age is doing. No one in this book is faking anything—the questions are real, the discussions are sincere and thoughtful and the contribution they have made to our community is undeniable.

Another reason I love this book is because it is organized around questions rather than answers. This is part of the work these writers have set for themselves because the questions they have asked, like all the important questions we ask ourselves, have no answers. That is not true only for adolescents, but for everyone. In case you haven't noticed, adults have not figured out how to control the human tendency towards violence, how to handle the complexities of emotional relationships or how to be true to ourselves in the face of peer pressure any better than teens have.

In my office, it's easy for me to get trapped in my own perspective, to forget how complicated everything can get. So I am especially happy to read such a variety of perspectives on some of life's questions. And I am proud to say that they are not teen perspectives on adolescent problems, but human perspectives on life's problems. This is not a school book or a class project or a teen book. It is a book as real as any you will ever read.

John Loonam
Director
The Institute for Media and Writing

Foreword

When youngsters enter school for the first time, they often do so with fear and trepidation because of the newness of it all. But as time passes, those fears and trepidations turn to joy and wonderment as the world of teaching and learning and reading and writing is uncovered. The child's imagination grows with every new experience, with every new exposure to the world of words. As time passes, that child begins to see and comprehend the world of words through both the eyes of the author and through his or her own experiences.

At this point in time, you are that child, putting together the lessons of the past to create this wonderful piece of literature. It brings together all those experiences in and out of school that make you who you are. Your work is a culmination of the creative juices that, in people, often times change the course of our world.

Congratulations to you, Ms. Quigley, and all of those persons involved in the evolution of the SPI project.

I know that this book represents the emerging storehouse of creative writing that is yet to be. I look forward to many volumes to follow.

John Angelet
Principal
Bayard Rustin Educational Complex

Introduction

"I hate cafeteria food!" Aaron cries out during fifth-period English. I've gotten used to the outbursts and my trite response since September has been, "Tell someone. Make a change," as if this will inspire my 10th grader to start a school-lunch revolution. Aaron just stares at me. His complaints speak to a greater dilemma in our young people: they are often unable to voice their opinion in an effective way that inspires real change. At the same time, they are afraid to listen to or consider perspectives that differ from their own (although Aaron might be hard-pressed to find an advocate for cafeteria food).

Young people aren't to blame. Despite the increasing power of the media and its ability to move information all over the world, there are many communities that consider it dangerous to oppose a commonly held view or to support those who do. Even in the United States, a country known for its diversity, the flood of information has overwhelmed our ability to consider different perspectives. Let's face it—it's often easier to turn to a cable channel or newspaper that reinforces our own judgments, our own truths, than to consider another point of view. The result of always having the "right" opinion is closed-mindedness and intolerance.

In response to this, the students at The Institute for Media and Writing created a book that celebrates different perspectives and asks the reader to consider the other side of things. They identified issues that touch them personally and posed tough questions that have no easy answers. Then they interviewed a diverse group of people and considered their markedly different perspectives. The result is not a question-and-answer book but a readable round table where everyone is given the chance to share and contemplate, without judgment, a different point of view. By publicizing these voices, we are valuing the living experiences that cultivate distinct perspectives. This is at the heart of teaching tolerance.

As for Aaron, he was able to develop his complaints about school into a consideration of a bigger question: *What prevents students from learning?* His focus on the connection between nutrition and education finally got people to pay attention and listen. Yenedy used her concern for a neglected friend to investigate the question: *Why do teens take on adult responsibilities?* Stephanie's personal struggle with a fellow student inspired her to contemplate: *How do you prevent conflicts with peers*? All of our students cared about their tough questions and challenged their own thinking to try to find some answers. We hope this comes through to our readers so that they, too, will publicize their own tough questions and acknowledge the multiple answers that these queries inspire.

The dynamic pieces of student writing in this publication reflect the collaboration of several dedicated, creative, and selfless educators. First, I would be remiss if I didn't thank John Angelet, Principal of the Bayard Rustin Educational Complex, and Dr. John Loonam, Director of The Institute for Media and Writing, for initiating this project after recognizing the opportunities that publication can offer students of all backgrounds. Tracy Faber, our 10th grade history teacher and my close friend, helped to map the original idea of our book and supported countless students during the entire publication process. Leila Sinclaire, our student teacher from Teachers College was unflagging in her efforts and, with the help of fellow student teacher Cynthia Moore, was instrumental in making this happen. Elizabeth Johnson, graduate researcher at Teachers College and "pop culture" guru, offered fresh ideas and new perspectives that enabled us to connect our project to the real world. Laura Starecheski, independent radio producer and our beloved teaching artist, spent hours interviewing and recording students to produce the attached compact disc. Finally, Kerry McKibbin of the Student Press Initiative at Teachers College, with the additional support of Leigh Reilly, facilitated every step of the publication process. Her wisdom, energy, and attention to detail made this daunting project feasible and rewarding for all involved.

The 10th grade students at The Institute for Media and Writing worked tirelessly over many months to create a text that reflects who they are and what they value. They were open to sharing their personal selves in their writing and were willing to listen to those whose opinions often differed from their own. They continually teach and inspire me every day.

Erin Quigley
English Language Arts Teacher
The Institute for Media and Writing
Bayard Rustin Educational Complex

Suggestions for using this book:

Right now, schools are responding to the call to develop the "whole" student by creating classes that address their personal needs. Whether it's called Advisory, Life Skills, or even Health, teachers take this opportunity to put aside traditional academics to focus on the "real world," the journey of adolescence into adulthood and all of the bumps that make it a challenge.

Our students created this book as an entry point for deeper discussion about that journey and added some helpful features we wouldn't want you to miss. In each chapter, after reading about the different perspectives on a central question, you'll find a list of student-generated questions that could lead to journaling or active class discussions.

Also, the students have generated a list of suggested resources that might help you, the teacher, create meaningful, memorable lessons around each question. Please know that the students had a fair amount of freedom to come up with the books, movies, music and websites that they thought best reflect their topic. Since the student's idea of what is "classroom appropriate" is not always the same as the teacher's, we encourage you to use your discretion when deciding which resources might supplement your class discussion! For example, showing R-rated movies requires parental permission. In the suggested resources, we have identified which movies are rated R.

Furthermore, we've included a compact disc so that our readers might listen to what our young writers have to say about the experience of exploring other perspectives. There is one track for each of the ten chapters. There are many possibilities here and we hope you will take advantage of them all.

On a final note, we want to make clear that we have changed the names of our young interviewees—even though they might not have wanted us to!—to protect their privacy. As you will see, our students were fearless in the topics they chose to tackle and they interviewed people who were equally brave in sharing their life experiences. Because of this, there are some questions that address sensitive subject matter.

Chapter 1

Teen Life

How do you avoid getting caught up in peer pressure?

We may not know it but people go through peer pressure all the time. You probably think you're the only person who's gone through it, but trust us you're not. In this chapter you will read about what to do and what not to do when you are being pressured. You'll hear from experts on the subject, and you'll also hear from regular, everyday people. No matter what you take from these different perspectives, always know that you're not alone.

Ashley Jagdhar

Ignore the negative; get caught up in the positive.

What is peer pressure? Peer pressure is when friends or other peers talk you into doing something you don't want to do. Peer pressure can be broken down into two categories: negative peer pressure and positive peer pressure. Negative peer pressure is being persuaded into doing something you don't want to do because your friends say you should. Positive peer pressure, on the other hand, is being pushed into doing something that you didn't have the courage to do, which might somehow influence your life and make you a better person.

Some research I found really influenced my own perspective on peer pressure. The behaviors that the following statistics represent are often a result of peer pressure. According to the Maryland Underage Drinking Coalition, the first use of alcohol begins around the age of thirteen. "Two-thirds of teenagers who drink buy their own alcohol, and junior and senior high school students drink 35% of all wine coolers sold in the United States." In a second study, the National Household Survey on Drug Abuse, conducted by the Department of Health and Human Services, demonstrated that, "Drug use is on the rise for twelve through seventeen year olds; since 1992 the smoking of marijuana has gone up 275%." A third study, performed by the American Lung Association, reported that, "At least 3.1 million adolescents and 25% of seventeen and eighteen year olds are current smokers." Finally, according to a report from Communities Responding to the Challenge of Adolescent Pregnancy Prevention, "Approximately 9% of 14-year-olds, 18% of 15 to 17-year-olds, and 22% of 18 to 19-year-olds experience a pregnancy each year." Obviously, peer pressure can lead to this sort of negative behavior.

"Peer pressure is a major issue among teenagers today," said Loleita Jagdhar, an adult who feels strongly about this topic because she is a mother—of me. "I think teens should be able to make decisions for themselves and not rely on others' opinions," she says. "When it comes down to pressuring others, no one should bully anyone else into doing something they don't want to do." When it comes down to it, though, at some point every one has been pressured into one thing or the other. Loleita says, "You just have to know when someone is pushing you too hard into doing something you don't want to do."

"Peer pressure at some point isn't so bad. Sometimes, people need an extra push to do something they didn't have the courage to do on their own," said my best friend. I had never really thought of peer pressure as something good before until I heard it from someone close to me. I guess not all things can be bad. I personally have experienced peer pressure but you have to be smart enough to know when you're being pushed into doing something bad. For example, in the eighth grade, someone tried pressuring me into cutting class. I thought I should do it because it would make me "fit in" but then realized what the consequences of my actions would be. If I had done so and cut class, I probably would have gotten in trouble and authorities would have gotten involved. That was a risk I was not willing to take.

Researching and interviewing someone who has a strong opinion on this topic has really changed my perspective on this whole thing. What impacted me the most was the change in perspective I got from interviewing Loleita. I never thought that peer pressure could have a positive side. As a teenager, I also didn't think that adults could relate to things like being peer pressured. Adults, to me, are supposed to be role models to children. You don't usually think your role models have been pressured into doing anything. Adults are supposed to always make decisions for themselves.

"Everyone has been pressured in their life," said Loleita. "Even adults, to this very day, get pressured into doing things they don't want to do. As a child, you're supposed to learn from a higher authority, or even from personal experience that no one can make you do

SUGGESTED RESOURCES

Books:

Bodega Dreams by Ernesto Quinonez

Always Running by Luis Rodriguez

Movies:

The Breakfast Club (R)

Mean Girls

Ferris Bueller's Day Off

Thirteen (R)

Basketball Diaries (R)

Cruel Intentions (R)

Internet:

http://uglydotcom.blogspot.com/

http://teenadvice.about.com/od/peerpressur1/Peer_Pressure.htm

http://www.kidshealth.org/teen/your_mind/problems/peer_pressure.html

http://www.cyh.com/HealthTopics/HealthTopicDetails.aspx?p=243&np=295&id=2184

anything you don't want to do. But remember, not all things are bad. If you feel you're being pressured into something good, like school work, it isn't to hurt you but to somehow push you to better yourself."

Sources

Bittman, Barry M.D. "Healthy Lifestyles in Teens: and the reality of peer pressure." Mind-Body Wellness Center. 1999. 8 March 2007. <http://www.mind-body.org/peer%20pressure.htm

"Peer Pressure The Good & the Bad". Teen Education Center Help .1996. 8 March 2007<http://library.thinkquest.org/3354/Resource_Center/Virtual_Library/Peer_Pressure/peer.htm

Birmania Paula

By facing your insecurities.

"Peer pressure is one of the toughest obstacles someone has to go through," said Jessica. She has been a close friend of mine for the past two years. Even though I know her very well, my interview with her was very nerve-wracking because, at the time, she was under a lot of stress from her own peer pressure situation. I was mostly afraid because she often keeps everything in, and then, when she's had enough, she cracks. Thankfully that didn't happen when I interviewed her. If it ever does, I'll be there to help her.

During our conversation, she reminded me that peer pressure is not helping me live or helping me solve any of my problems, so why should I please others? "I imagine it would be boring if there weren't any problems [with our friends pressuring us], but there has to be a balance," Jessica said.

I take it into consideration because I have been peer pressured before, a lot. I've been asked to go to hooky parties. I also was pressured a lot to be intimate with someone my freshman year. I find it disgusting and sickening that I got caught up in it. It's amazing that some people actually give into this just to become "cool." The people doing the pressuring can also be cruel. They will get others to do certain things they wouldn't do on their own, and then the people pressuring will drop them if they opt out of the pressure.

I first started seeing peer pressure when I started junior high school. One of my best friends since elementary school started giving into it. He started to do drugs, drink and he eventually joined a gang in eighth grade. The only reason he did it was because he was being pressured by his "friends" and his brothers at home. I really thought that he was smarter than that and he wouldn't give in.

Jessica seems to be getting wrapped up in the same thing. "I am almost certain that I won't get caught," she said to me during the interview and it made me get closer to her as a friend. If there were no peer pressure, wouldn't it be a better place for teens and everyone else? I know it would certainly be a better place for me than it is now. I wouldn't have to worry about my friends telling me what to do all the time.

Is there any way that pressure can be avoided? According to Jessica, the best way to stay true to yourself is to make sure that you're doing that because you want to and not to be envied. I have my own suggestions for avoiding peer pressure. Teens should remember that they should just respect themselves, think about what they're doing, and think twice before they do or say something they might regret the next day. Whenever one of my friends pressures me to do something, I always seem to remember one thing: I am only going to be in high school for two more years and then after that I'm never going to see these people again, so why should I make myself into something I'm not and embarrass myself now and then regret it later? Of course it's easier said than done, but I'm not going to give in easily; I'm smarter than that. As Jessica once said, "Many times in your life people will try to make you into who they want, but it's only you who can make that the person you want to be."

Crizeydi Pineda

Learn from the experiences of others.

Once my friend told me to cut school with her and go to her house. I did it expecting it to be harmless fun between us two because we were like best friends. Little did I know she had other plans. We were not alone. When we went to her house, there was a group of guys waiting in front of her building. When we reached the building, the guys followed us inside making me hesitate before going in. My friend noticed my hesitation and she told me, "It's all good. They're with me." I am still scared when I think back because there were at least five of them there. It made me feel so uncomfortable. I went into her house because I did not want to look like a coward in front of her. I wanted her to think I was cool. In other words, I was trying to be something I was not in order to fit in.

We hung out for a while and we drank some of her parents' expensive liquor. The guys must have gotten sick of waiting and the "leader" of the gang was like "what the f***? We didn't come here just to stare at each other." The rest of the story is that we got really drunk and got into a lot of trouble. I chose my friends wisely after that and did not follow anyone anymore. The experience taught me you cannot trust anybody but yourself. I still talk to the girl to be nice and share hellos and goodbyes.

"Of course, if they're a family member I have to defend them.... with teeth and claws," said Roberta, a Hispanic 10th grader who has experienced peer pressure. Roberta is trying to forget about her past peer pressure experiences. She is now trying to avoid peer pressure so she can reach her future goals. She experienced peer pressure when she was in elementary school, junior high, and now recently in high school. She can recall her experiences as if they occurred yesterday, but she wishes it never happened and that tomorrow she will wake up and it will all be a dream. As I look at her, I notice how sad she seems. It's as if it took a lot for her to recall this story because she is trying so hard to forget it. I do not want to talk about it either, but it is also good to let things like this out and tell the world so at least a few people will be helped.

Roberta grew up in East Harlem, which isn't really a bad area. She knows people who have been peer pressured also. In elementary school, a girl picked on her by pulling her hair, putting glue on her book bag, and poking holes in her jacket. This occurred, at first, because she was peer pressured into doing things for that girl. Then Roberta finally stepped up to her and the girl did these things. Roberta got sick of her behavior and pulled her hair right back until they ended up fighting. The girl tried to apologize the next day, but Roberta did not want to hear it. Roberta felt that if she gave in to the girl, she would do the same thing back to her. Therefore, by not accepting her apology, she avoided this from happening again. Roberta caught up with that same girl in junior high, but the girl felt that she would have held a grudge against her so she avoided her.

Recently in ninth grade, a different girl she hung out with started to peer pressure Roberta as well. She would make her cut classes and go to parties during school. She tried to tell the girl she did not want to but the girl made it clear that if she did not, she would not talk to her or be her friend. So once again, being afraid, she did what she was told. I feel that Roberta did the right thing by fighting this girl. I know that violence is not the key to solving problems, but at least she tried to stop it. She did do a good job by avoiding that girl.

Our experiences are related because we were both peer pressured into doing things we did not want to. Then by putting an end to it, we were tortured horribly. Learn from our personal experiences and don't be afraid to say no. Do not be afraid to tell a teacher, a parent or a guidance counselor about your problems with peer pressure if the person continues to bother you. It will really help you if you learn from our mistakes, not your own.

Celene Santiago

It's your choice.

Not many people know that peer pressure is a common experience people go through. Grown-ups go through it all over the world, but the most vulnerable people are teenagers. Here's the question: how do you avoid it?

Teenagers always go through peer pressure, whether positive or negative. Therapist and social worker Nicole says, "Peer pressure can be good—good people, good pressure—and it can be bad—doing things you know are wrong with bad people. You just have to be strong enough to pull away and choose the right path."

Being a teenager myself, I have also had problems choosing the right path. When problems like peer pressure come along, it's always good to go to family or an adult for help. "When I didn't know what to do," Nicole says. "I went to my mother and grandmother for help. I also turned to teachers for help."

It's very hard to realize when you are being pressured. Sometimes you don't notice until someone points it out to you. When I was in eighth grade, I didn't realize I was being pressured. I was skipping classes with my friend for about a month. It wasn't all my classes, but I did it a lot. The first time he asked me to skip, I said no and told him it was wrong to skip classes just to eat at McDonald's. He kept telling me that it would be okay and that no one would find out. I still said no. Then he started to tell me that I was a goody two-shoes for not wanting to skip class. That got me really angry, so I gave in. After about two weeks, I began asking him to come skip classes with me. This continued for about a month.

One day I came home from school and my mother told me that she got a call from school. I already knew what the call was for. She told me that it wasn't like me to skip class. I told her that it was time for me to stop being the goody two-shoes and live on the "wild side." She told me that I had to open my eyes and notice what was really happening to me. It took awhile, but I finally figured out what my mother was saying: I was being pressured by my friend to do something I knew was wrong, but didn't notice because I was blinded by trying to impress him.

I stopped skipping classes and told my friend that it was wrong for him to pressure me and even more wrong that I listened to him. I began making my own choices, and found that I was happier. Yeah, I did end up losing my friend, but I learned to be strong and confident. "Who do you choose to be faithful to?" Nicole reinforced. "Your friends or you?" That time, I chose me.

If I knew then what I know now, I definitely would have been happier and life would have been easier to handle. I have to admit; not knowing what to do made me feel alone and out of place. Hearing Nicole say, "You're not alone. There are others out there just like you," made me feel that I made the right choice.

Going through negative peer pressure is a normal thing in life. Learning how to deal with it is the most important thing because it teaches us how to grow and change. "Adults don't always have the answers," said Nicole. "And neither do students. It's up to you to make the right choice." Going by what you think is right is the best way to avoid peer pressure. Talking to someone is great for advice, but the answer to the problem really comes from you. Your instinct is usually the right answer so go by it. All someone else can do is give us a little push, and let us do the rest.

Discussion Questions

1. Have you ever caught yourself pressuring others to do something they didn't want to do?
2. Describe a time you felt peer pressured.
3. Why do you think people put pressure on others?
4. How can we resist peer pressure?
5. Can peer pressure ever be positive?

What influences how cliques are formed?

In every school, students are unconsciously negotiating the boundaries of friendship. Some have created tight bonds with their classmates. These students are in cliques. You might be in one and not even realize it. Cliques are one of the things that every teen deals with in some way, but most teens are unwilling to talk about it because of its sensitive and ephemeral nature: groups and the people within them are always changing and evolving. If you're in one and you don't acknowledge or discuss it, you're left to figure cliques out on your own. In this chapter, we tested the issues about the topic. We formed a group and interviewed different people that have a range of perspectives about what influences how cliques are formed. They gave many insights that reinforce the idea that there's no simple answer to our challenging question.

Sandra Cordero

Strong bonds hold cliques together.

Diana is a social teenager at Bayard Rustin Educational Complex and to her there is nothing wrong with having a close group of friends. Diana has a great personality. She is open-minded and not shy at all. She is also confident in herself and can easily make friends, yet she prefers to stick by a select few. You might refer to her preference of friends as clique-y, but don't we all have a group of friends that we hold dear? How are these cliques created? What are they based upon? Why are our ties with our friends identified as cliques?

Diana thinks there is nothing wrong with cliques. She believes, "That is how you know who your real friends are, because they stick by you." In other words, as Diana goes on to say, "Close friends are good to have because they have each other's backs." Though this may be true, Diana also thinks that there are some cases where teens can take advantage of the strengths of their group of friends. "Sometimes they think they have a lot of power because they have a lot of people with them. They hurt others." Surely, it is wrong to hurt others with the help of your friends yet in a peculiar way it also shows how strong the bond is. "Personal opinions don't matter when it comes to helping your friends," Diana believes. Without trust and loyalty, the bond we have with our friends wouldn't be so strong.

Cliques used to be a way of survival in school. Friends brought out the confidence in each other and stood together. To me, they where like a comfortable and safe way to socialize in school. Instead of going around, meeting other students and trying to make new friends, I would just stay with the same people. It made me feel more secure since I was already used to those people. I felt like I didn't have to try as hard to fit in with my friends since we were basically the only friends each other had. We felt like we could not afford to lose each other.

Unfortunately, we did not feel the same way about other kids at school. Those who tried to approach us would either be sneered at or told off. Among my friends, we would criticize other kids simply because they weren't a member of our clique. There were times when someone wanted to befriend us, but we would usually be skeptical of them and give them a hard time. We felt like no one else was like us and anyone who wanted to be friends with us was just trying to pretend they were one of us. If they weren't skinny with long hair and liked specific things—for example, the color blue or specific TV shows—we would make fun of them and, on rare occasions, would even make them cry. By doing this, we put them down and sometimes crushed their self-esteem which, now I see, is unfair and nasty. To us, it seemed so easy and harmless to talk about people in front of them or behind their backs. Since we all did it, it seemed only natural to us. We only felt sure of ourselves when we had each other. Alone, we would be shy and hardly noticed, quite like the kids who wanted to be friends with us.

To me, cliques used to be like a safety net for a trapeze artist—I could mess up, but they would always catch me. Having gotten used to my rather small group of friends, spending even a day alone or with another crowd was somehow overwhelming and was disapproved of by my clique members. My opinion was very simple: once you're accepted into our clique, you stick by us and we'll stick by you and there was nothing wrong with that.

Realizing how unfair and inconsiderate my friends and I were, I'm grateful that I always have my friends by my side, holding that trapeze net and letting me feel safe. But as good as it feels to be safe and secure, it feels even better to know that I am also able to help others feel secure and confident instead of putting them down. Now I hold the net.

SUGGESTED RESOURCES

Books:

On the Fringe by Chris Crutcher

Cliques by Heather Moehn

Life and Loss: A Guide to Help Grieving Children by Linda Goldman

The Seven Habits of Highly Effective Teens by Sean Covey

The Complete Idiot's Guide to Friendships for Teens by Erica Lutz

Magazines:

Cosmo Girl

Teen Vogue

Seventeen

Movies:

Mean Girls

The Sisterhood of the Traveling Pants

A Walk to Remember

The Wood (R)

TV shows:

Friends

Juvies

Music:

"Welcome to My Life" by Simple Plan

Glenys Rodriguez

It's all about choices.

I look around the hallways and the cafeteria and I identify students grouped together. What does that mean? Sometimes I wonder why when we get used to certain groups of people, we stick together and these groups become cliques. I had a chat with an assistant principal, Ms. Polsonetti. We carefully discussed why most students form groups. Ms. Polsonetti and I did not come up with an answer but we thought of a few reasons. After our discussion I put more mind to the situation and came up with some other ideas.

Cliques are a way to determine different groups and are often formed in junior high. They sometimes form because of what the peers have in common or because of stereotypes. Certain races or genders will stick together because that makes them more comfortable. Most groups form because they look at it as a way of categorizing their peers. The most well-known cliques are the jocks, the popular students, geeks, and the bullies. Even though some cliques are not even labeled, we define cliques as a group of people that is together because of a purpose, like a sport or some type of club. Peers sometimes just do not notice that they are in a group. As Ms. Polsonetti says, "It depends on the person. Sometimes they join groups to be accepted or to have a core group of friends."

I did join different groups until I found the right group of friends that I knew I would get along with and have fun with. It sometimes takes a long time to find the perfect clique to feel comfortable with and it is sometimes necessary to try different cliques. Changing groups is a good way to find out what you are interested in. Depending on what clique we choose to be in, there might be a positive or a negative outcome.

When it comes to peer pressure, sometimes we need to be careful. In junior high, I knew a girl who started hanging out with the wrong group of girls who smoked and did not give importance to school. She may have started to hang out with them because she thought of it as being cool—these girls were known throughout the school. 'Til this day, I don't understand her motive for becoming lost and disobedient. Later I found out that she was pregnant. Before she started to hang out with those girls she was a sober and caring person. After, she became a careless, alienated person because of the bad influence the group of girls had on her.

Last year as a freshman I did not know with whom I should hang out. Carefully, a group of girls, including myself, started to form a group until we were inseparable for a period of time. As our friendship grew, so did the drama. Some of us left this old clique and started a new one with the girls we trusted. Others may not find us together because we get together mostly at lunch and when we plan to go out. The reason our clique formed was because we found a group with which we had something in common and with whom we felt comfortable. Sometimes we have confrontations, but we are mature enough not to take it to another level. You may not find that in most cliques.

When deciding to join or form a clique, first decide what you want. Often teenagers don't know why they join certain groups. Maybe teens would be happier if they answered our central question, what influences how cliques are formed? Talking to Ms. Polsonetti made me realize that you get to learn many things about yourself in a clique. I am a person who enjoys being in a clique and that is something I choose to do. As Ms. Polsonetti explained, joining a clique and staying with them "is all about choices, whether it is the right choice or not." I know my choice to join a clique was the right one.

Christopher Showalter

Deciding to take the road alone.

A group of men trot down the street and cops look them up and down as if they are common criminals, but really each of them is a unique human being.

My momma told me not to judge a book by its cover, but in a conversation with my best friend, Shawn, it wasn't hard to figure out why most people look at cliques like they have a hidden agenda. Shawn stated, "A clique is a group of five or fewer friends that have each other's back in a tight relationship. A clique is formed when a group of friends have a tight bond with one another creating an unbreakable friendship." As an outsider looking in, Shawn states, "I look at it as a gang of good friends, but smaller."

One experience changed Shawn's perspective forever. When Shawn was young, one of his friends was beaten up because he lied about joining a gang. That experience shocked Shawn because he always compared cliques to gangs. He noticed that, like gangs, cliques often were surrounded by friends and wore similar clothes. Also, the public's view of cliques is negative and this affected Shawn's opinion of them. After that incident and further thinking about cliques' reputations, Shawn was set on being independent and focused on his life ahead without a close group of friends. He believed we can all be happy and reach our goals if we stay out of a clique.

Through the years of my life, my view of cliques has been different from others. They are no more than a chain, a tight group. True, my mother regards cliques as gangs, like Shawn does. Parents and other judgmental adults tend to wonder why their children are always hanging with them, but everyone has their own opinion and I feel completely differently. I believe they provide a family-like retreat or a fight club that has your back. They can expand and attract people, not necessarily exclude them. To me, it's also about hanging out all the time with a group of friends.

It is true that there are negatives to every positive. Sometimes I see a kid join a group that he feels will make him more comfortable. If, however, he is separated from this group, he feels lost or unable to adjust to life on his own as an individual, like a lost sheep. You can never find that perfect group, so you better be able to survive on your own. Through the years I sort of see what Shawn sees: cliques can be dangerous. They can bully individuals who don't have their own group and they feel invincible because they roll in numbers. Instead of joining the crowd, Shawn has decided to take the road alone. I admire him for his bravery but as one of a clique myself, I know you can take company with you on the road. Shawn's face glows, as he stands by himself in a clique-less life where he knows he is responsible for all of his successes.

Discussion Questions

1. Why do people feel the need to form or join cliques?
2. What makes a clique become strong?
3. What makes a clique different from a gang?
4. What benefits do you receive from being in a clique?
5. What are the positive and negative effects of cliques?

Why do people feel the need to impress others?

Our dress is our identity. Through our dress we live, move and express our social being. You can change the way you dress, your personality, your hair and everything about yourself, but does that make you a better person? Well, that's what we wanted to know. Why change yourself or try to impress someone else to fit in? Why do teens always want to fit in? Are you that unhappy with yourself? Why not just be yourself? Everyone has their reasons for doing the things they do and we were very excited to find out what they were. We each interviewed people with different perspectives on this topic. So imagine, if they were all in one room together and we asked them all the same question, they would all bump heads because everyone has their own opinion. Surprisingly, everyone was straight-to-the-point and totally honest for such a delicate topic. Will you be?

Cesar Balbuena

People change their identities so that they can be liked.

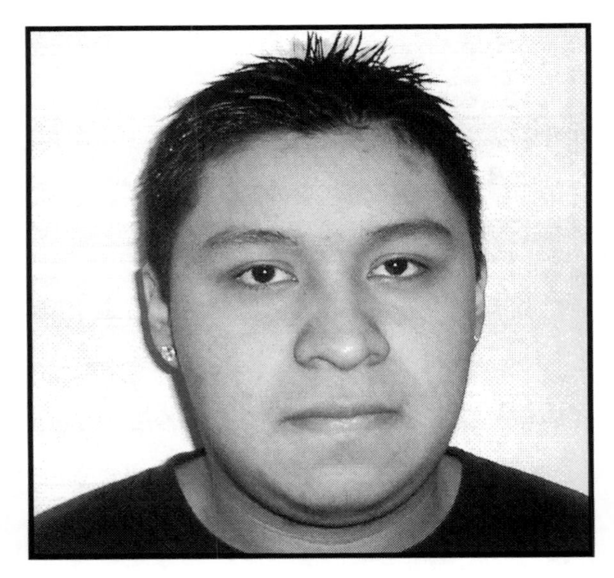

I once felt the need to impress a girl who I liked and I thought that if I changed the way I dressed she would look at me in a different way. I was 13 at the time and I wanted to look and act more mature and older than my age. I felt that I loved her so much that I even tried to change the way I talked. I didn't just do it

SUGGESTED RESOURCES

Books:

Gossip Girls by Cecily von Ziegesar

Origami With Dollar Bills by Duy Nguyen

Dress to Impress by William J. F. Keenan

GirlWise: How to Be Confident, Capable, Cool and in Control by Julia Devillers

Magazines:

Seventeen

Teen Vogue

Cosmo Girl

People

Movies:

The Breakfast Club (R)

Pretty in Pink

Sixteen Candles

Clueless

Heathers

She's All That

Tart

Thirteen (R)

Mean Girls

TV shows:

Dance Life

Parental Control

Room Raters

My Super Sweet Sixteen

Ugly Betty

for her; I did it for myself. It made me feel that I was changing for the good. Now, I don't recommend changing for the person you like because he or she and others might look at you as if you are weird.

I had my own experiences, but I wanted to know more about it from a girl's perspective. Jessica is a 15-year-old friend of mine. We have known each other for two years and I consider her to be a really pleasant person. I noticed that she always shows her true side and who she really is. For example, she dresses like a girly-girl because she is a girly-girl. Her outside matches her inside. But she wasn't always like this.

Jessica told me that she once tried to impress a boy because she liked him, but he didn't like her. She thought that if she changed the way she looked the boy would pay attention to her. And that's what happened. After she made some changes to the way she dressed, the boy started to like her. Obviously this made her happy. I was there when she started to change and I felt like she was another person. I realized that her attitude was different too and also the way she looked at people. I thought that she had completely changed and that she was no longer the person who I had become friends with. They went out for five months and then broke up because he didn't like the way she was acting. He didn't like the fact that she had changed to be with him. After they broke up, she was her usual self again.

We're still friends, as close as we were before. What I learned from the experiences that Jessica and I had is that if you are yourself, you will find the right person for you. There's no need to change something for the person you like. Changing the way I dressed for the girl I liked didn't work because she didn't pay attention. She said that it was too late; that I should have changed the way I dressed before because by that point, she liked someone else. What I realized is that there are many more girls out there besides her and I learned my lesson: never change something you don't like about yourself, unless it's going to make you truly happy. Now I'm glad that I didn't go overboard to change myself for that girl because I've moved on and have a girl that likes me the way I am. I don't have to change nothing for her. And I feel good about my life just the way it is.

Shanell Holback

I can't be me.

Teens suffer from all kinds of pressure. Some teens feel that they need others to like them so they can move on in life. I spoke to a teenage friend of mine who is going through the experience of trying to impress others so that they have a positive view of her and she can succeed in life more easily. She chose to keep her name a secret but she clearly suffers from the pressure that a lot of teens face.

I chose to interview this young lady because when I told her about the topic of teenagers trying to impress others she told me she could give me a lot of information. We have been friends for two years and I knew that she would be the perfect person for this interview. When I spoke with her, she told me that she tries to make people accept her so that she can get more from life. For example, she feels the need to impress her boyfriend so that he won't leave her for another girl.

I wouldn't try to impress anyone other than my mother and God because they created me and they help me live another day in the world. I haven't struggled much with the pressure to impress others because I always felt people should like me for me. I don't have to make anybody like me in order for me to like myself.

It was interesting for me to interview this young lady about what she is going through because I have a very positive view of her. I can tell that she's a very nice young lady with an open mind but it seems as if she's using it in the wrong way. I think people who feel that they need to impress others don't really feel that good about themselves, so they feel better when other people like them. The thing that many people fail to notice is that people don't like you for you when you put on a front for them. They're liking a person that you are not.

As the interview ended with this young lady, I began to think about the things she shared with me. In the interview, I asked her if she thought she could impress others without trying and she told me, "Yes, because it is in my nature to impress others." However, when I asked her if she went outside herself to impress others she said, "Yes, because I want people to think I'm better than I really am." In my eyes, both questions are asking the same thing but in different words. But as you can see, she gave me two different answers. That might mean that she's not aware of how important it is to just be herself. She told me, "If a person says something nega-

Music:

SUGGESTED RESOURCES

"How to Touch a Girl" by JoJo

"Wouldn't Get Far" by The Game

"Sorry 2004" by Ruben Stoddard

"Unpretty" by TLC

"Exposed" by Chante Moore

Websites:

www.myspace.com

www.sconex.com

tive about me, they don't know me well enough to judge me but if they have a positive view of me, then I would tell them that they should keep hanging out with me and they might learn something." With this in mind, I leave all teens to think about this: if you're trying to impress other people, you most likely haven't taken the time to impress yourself. Your own understanding of yourself is more important than anyone else's view of you. When you understand yourself, people will begin to understand you.

Katherine Imbert

Fitting in and fitting out.

I know a girl who takes a long time to get ready in the morning, looking for the perfect outfit. She says that she isn't really trying to impress others, but that she just wants to look cute. She doesn't ever want to look ugly because she says it would mess up her reputation. She is popular and known as the "best dressed" girl in school. I always wonder why she does this. I suspect that she really is trying to impress others but she doesn't want to admit to it. Maybe she's saying the truth.

My cousin, Maria, is 16 years old. She attends Marie Curie High School and liked this popular boy. She went up to the guy and began talking to him. He asked her, "Why do you dress like that?" She said, "I dress the way I want to dress because I don't have to impress people. I'm just me. I really like you but I can't change myself for you." Ever since that day, they've been going out. She refused to change herself for him and he respected that. According to Maria, impressing others is a waste of time. This story proves that it's not worth changing yourself just to impress someone else because they could like you just the way you are. Maybe you just don't know it.

My other cousin, Dalila, also agrees that impressing others is a waste of time. She says, "I see no point in impressing others because they will never be satisfied. Others will always have something to say no matter what." She attends Washington Irving High School and knows mostly everyone. She sees a lot of people dressing up like they're going out on the runway at a fashion show. She estimates that 60% of those girls are just looking for attention, and the other 40% are just being themselves. My cousin says, "It's a waste of time because you try so hard to look your best and people still humiliate you behind your back. The best idea is to impress yourself and no one else because if you're not happy, chances are no one is going to be."

There are a couple of students at my school who like to dress to impress. Personally, I believe I am myself and if you don't like what you see then don't look at me. Sometimes impressing others can cause problems because you might impress someone who already has a girlfriend. Then you're going to have problems with a boy just because you weren't being yourself. I recommend to teens that they think twice before they dress up just to impress.

Ronniece Parris

Because they just want to fit in.

Misha is 16 years old and attends a large school in Pleasantville, New Jersey. She lives with her mom, dad, and older brother. Her motto is, "Life's too short to walk around unnoticed, so why not do what you want?" Misha gets As and Bs and is a really sweet girl, but she feels that she has to impress others in order to feel wanted. Misha worries she's not going to fit in so she does stuff like changing her appearance and her personality. That doesn't mean she's a fake person, it just means she hasn't found herself yet. She doesn't know if she wants to be herself or if she wants to act like someone else.

Misha is a really good friend but sometimes I feel her poor decisions can lead to bad things. I will never forget the time when she started ignoring and leaving her friends for some boy. The first time she met Johnny, from what she told me, he seemed like a pretty nice boy, but then they started getting really serious too fast. I stopped seeing her; I didn't talk to her on the phone anymore or anything. At first I didn't mind because I was really happy for her. Then I realized how much I missed my friend. Eventually, he broke up with her and she was crushed because she really liked the guy. I felt bad for her because she lost a few of her closest friends in the process, but I kept saying in my head, "I really don't care."

Later on I felt sorry for her because I wasn't being a very good friend. I said things to her like, "That's your problem, not mine." I knew it hurt her but I didn't care. I wanted her to feel the pain I felt when she wasn't there for me, when she was too busy with Johnny. Later she confided in me that Johnny tried to change everything about her. He didn't like her weight or certain clothes she wore or her make-up. The worst thing was that she would try to change for him. Now that I look back I feel guilty. I think I should have been there for my friend in her time of need.

All I can think of when I hear that people are changing for others is, "What's wrong with you?" If someone is your friend or if someone truly likes you, they would like you for who you are; no matter what you wear, how you have your hair or anything else. So think about that the next time you are trying to impress someone, because in the end if you can't impress yourself, then you can't impress anyone.

Discussion Questions:

1. What are the consequences of impressing others?
2. Talk about a time when you found yourself wanting to impress someone? Why did you do it?
3. What would you do if you had a friend who completely changed for someone else?
4. When is trying to impress others valuable or necessary?
5. When you are trying to impress someone, do you think you're being yourself?

Chapter 2

Relationships

How do you know when it's the right time to have sex?

There are a lot of things going on in the mind of a teenager, and sex happens to be one of them. There are a lot of sexual influences and pressures on teens; they often end up thinking about sex all the time. People have very different opinions on the topic. How do we really know when it's the right time to have sex? We hope that hearing someone else's perspective will help you become more comfortable with your own decision.

Diana Gonzalez

It's better to be safe than sorry.

Ideally sex is something you do with your partner to demonstrate your love for each other. Often it can be for pleasure alone. Either way, sex is a complicated issue. Depending on when or with whom you want to do it, it is your own decision. Sometimes teenagers feel pressure from their partners to show their love for them. There are many definitions of what sex is. However, if you read books, ask a person, or if you feel comfortable asking your doctor, everything will be clear. In my opinion, sex is something special and natural that humans experience.

We all know the horror stories of friends whose sexual encounters become nightmares. One time a close friend was having sex with her boyfriend. She was in love, and they were not using protection. When she went to the hospital the doctor told her she had chlamydia. Chlamydia is the most common sexually transmitted disease, but chlamydia goes away with treatment. My friend was scared at that moment, but once they told her that it could be treated, she was fine. Still, this made her think twice before being intimate with just anyone.

I talked to my 17-year-old sister, Maria, about when we will know it is the right time to have sex. She is not currently dating anyone. She would like to have sex, but she is waiting for the right person, somebody who she loves and who loves her, someone who can communicate and loves being together with her. Her perspective on the topic is that when you have sex the best way to protect yourself is by using condoms. "Sometimes things just happen and you don't plan them." She thinks age does not matter, but your mind will tell you if you are prepared. Sex can be dangerous because you can get an infection, a deadly disease, or fall pregnant, unless you protect and respect yourself. She developed this perspective from all the reading she has done on the subject and what she studied in school. The only real people who give her insight on the world of sex are her friends. Our parents consider it a touchy subject and they don't ever discuss the details of sex with us. All we know is that we should wait until we are married.

My perspective about this topic is really like my sister's. I think the right time to have sex is when you feel completely ready and prepared. Feeling prepared for me is when you do not feel peer pressured, when you love that person, and when you feel comfortable with him or her. You really have to think about who you are doing it with, and, most importantly, that you would not regret it afterwards. You also have to protect yourself so you don't get a disease or become pregnant.

It is an important topic because sometimes there are many teenagers in this situation who don't know what to do or who to talk to. My sister and I both agree that when you have sex, everything changes in your life. You are not a teenager anymore; you are an adult and you could become a parent if you are not careful. I am lucky to have such a good influence and I appreciate that she is setting a good example.

SUGGESTED RESOURCES

Books:

The Naked Truth About Sex: A Guide to Intelligent Sexual Choices for Teenagers and Twentysomethings by Roger W. Libby

Sex in School: Canadian Education and Sexual Regulation edited by Susan Prentice

Movies:

Thirteen (R)

Internet:

www.mundofred.com

www.kidshelp.org

www.allposters.com (sex education)

Guest Speakers:

Planned parenthood representatives

Celenny Lantigua

When it's true love.

Is abstinence the best choice for teenagers? My friend, Yahaira, has an opinion on this tough question. She is a Dominican 17- year-old girl who, most of her life, grew up in New York. She's really pretty and you would think that by now she would have a lot of experience with sex, but she hasn't. I admire her because she has stuck to her beliefs even when it was difficult. Yahaira believes, "Once you start it, you can't stop it." To her, abstinence is the best choice for teenagers. Yahaira hasn't had an opportunity because she thinks she hasn't found the right guy.

This doesn't mean she thinks that sex is going to kill her or hurt her in any way. And when it comes to waiting to have sex until marriage, she thinks that's unrealistic. Instead, she believes people should wait until they feel completely prepared and have found the right person and when they're sure they won't regret anything it leads to.

Pressure to have sex is really common these days. Luckily Yahaira hasn't been pressured that often and her friends respect every decision she makes. There have been occasions when she could have had sex but she chose not to because she thinks that the guys who gave her the opportunities were not worth it. Sex has turned into a necessity for some people, especially the people who have done it already. She says, "Teenagers that have chosen to have sex at an early age made the right decision because if they do it is because they want to and because they feel prepared." Most people fall in love with the person that they first have sex with mostly because is the first person to see you without clothes and be intimate with you.

My experience is a bit different from Yahaira's. A lot of my friends are having sex now. It seems almost normal in high school. Some of my friends have been having sex since junior high. They can't resist it, being surrounded by so many images of sex. Whether it's music videos, movies, television, or even teen magazines, there's a message right now that everyone should be having sex. You almost feel like an outcast if you aren't.

Even though it's easy to get caught up in it, I think teenagers are taking things too fast. The truth is, not everyone is actually having sex. A lot of people lie about being sexually active because they want to fit in. This leads other people to consider doing it because they think their peers are telling the truth. Meanwhile, our parents are always telling us to wait because they don't want us to make a mistake, like acquiring an STD or getting pregnant. Like always, we ignore what they tell us. Once I used to think that by having sex with your partner you would prove your love for them. Now I know a lot of people do it for pleasure. For me sex is something important and I would prefer to do it with the person I choose to love. If I save it for those rare occasions, sex will be a positive experience, not a negative one.

Careema Parks

After marriage, you can have all the sex you want.

Many high school students believe that kids in high school should not have sex, with or without protection against STDs and pregnancy. Although teens may believe that sex at a young age is bad, it is rare that you will find a teen who practices that belief. People always say "kids today this" and "kids today that," so I thought it might be interesting to see what a woman who isn't from our generation had to say about the matter of when it is the "right time."

Deborah, my aunt and godmother, is a mother of five and has a rather interesting way of looking at it. She is a very outspoken person, but she has a sugary-sweet way of discussing the most awkward topics. I met with her in her apartment after school and she was quite ready to discuss her perspective on choosing the right time to have sex. She started having sex at a fairly young age, around thirteen or fourteen, and she admits that her sexuality is important to her. She believes that you should wait until after you are married and then you can have all the sex you want. She didn't wait until she was married, but she doesn't regret her decision. Her first time was with a cute Spanish boy named Eduardo. Eduardo wanted to marry her and she thought that he was just saying that so she would have sex with him. They stayed together for awhile after their first time together, but like most childhood relationships, it didn't last very long. Although she didn't wait she still believes that her children should wait because she wants her children to make better decisions than she did.

As far as when parents should tell their children, my aunt is very clear: "I think kids should learn about it in school. I would only talk about it if they wanted to talk about it." Her parents didn't talk to her about sex. Maybe if they did she would have made the decision to wait. I think that more people should talk to their kids about sex. If they did, they would have a more relatable way of looking at sex. School teaches kids about sex but they teach them the textbook stuff. Also they teach them about STDs and all the bad stuff that can happen. If the parents gave them their perspective and shared their encounters with sex as teens, maybe it would reduce the rate at which teens are having sex.

Believing that sex is so fun and enjoyable might play a part in why so many teens are choosing to have sex at such a young age. But there are other reasons teens are not waiting for the right person and just having sex with the "right-now" person. For example, many teens feel pressured to have sex and be like everybody else. I have a friend who even lies and says he is having sex so that he can be "down." It's as if he's not a man unless he has sex. The reality is it takes more than sex to be a grown-up. Sometimes the most grown-up decision you can make is to wait. Also, many teens who do find the right person don't understand that sex is not necessary for a relationship to work. You can still have a boyfriend or girlfriend and not have sex with him or her.

I feel lucky because most of the people in my life are pressuring me *not* to have sex. They care about my life and my future and that has made a difference. I know that everyone is abstinent for different reasons. Mine is simply that I haven't found the right person. Some people say that they are happy with their choices, like Deborah. Some people are not so happy. Whatever the reason someone might or might not choose to have sex, I think you should wait for the right time and right person. How do you know it is the right time and person? When I asked Deborah, she said, "You will just know."

Marsia Rochelin

There is no such thing as the right time.

While the topic of sex brings a lot to mind, abstinence brings even more to mine, since I am a virgin. I honestly have trouble knowing when it's the right time to have sex. Many times I felt like caving in under the pressure of forever being called the "virgin." I was once in the hot spot, being pressured for sex by a childhood friend of mine. Even in the heat of the moment I didn't do it. It wasn't the right person, place, or time.

From the time I was a preteen, sex was always the topic of conversation with my friends, my sister's friends, their friends, my brothers, their friends, and everyone else. In junior high, it felt like everyone I knew, one by one, started having sex, which made me feel left out and made me grow curious. I wanted to fit in and not feel left out. I felt so pressured, but I'd tell my sister about my real thoughts. She'd give me the rundown on the way boys thought, and how having sex would effect and change my life, and mention STDs and teen pregnancy. She did for me what no one had done for her and for that I am forever grateful.

I interviewed my friend Jasmine and got the perspective of a young adult who believes that there is no such thing as a right time because, according to her, we are in a generation that is fast. In her eyes there is no such thing as getting to know each other and moving at a steady pace anymore.

Everyone around her was having sex—the popular kids, jocks, the steppers, the kids no one would expect, like the quiet ones. It seemed like all of them were doing it. She lied to everyone about being a virgin. She didn't want to be left out and truly just wanted to fit in. She actually lost her virginity three days before her fifteenth birthday. It wasn't with the right person. In fact, it was with her boyfriend's friend. "He was extremely aggressive and out of my league at the time," she told me when we spoke. She lied to him that she had three sexual partners prior to him. He was moving extremely fast and rough. She admitted, "I wasn't being used because I was using him, too."

Jasmine believes that there is no such thing as a right time. She says that she doesn't regret anything because everything made her who she is now and also life is too short to be living in regret. Her story made me unafraid to ask about sex. I've realized that I can learn through the experiences of everyone else around me. Then I can make sure my experiences suit me and not others. That way I make sure I am pleasing Marsia and not trying to please others. With this approach, I fully understand what I am doing and what I, alone, can endure, especially when it comes to sex.

Discussion Questions

1. How do you know it is the right time to have sex?
2. What age do you think is appropriate to have sex?
3. When is the right time for parents to talk to their kids about sex?
4. What role does the media play in teenagers' sexual choices?
5. How much influence does a person's religious values have on his or her sexual choices?

How do you make a romantic relationship work?

How do you make a teen relationship work? Well, there are a lot of answers to this question. This is a very common question asked among teens, but has it ever been really answered? The answer depends upon your unique perspective and situation. Too many teens go through relationships—and have ups and downs—without talking to each other about how to make it work. If you don't learn about relationships now, you will have to learn about them later as an adult, perhaps in more difficult situations. Teens are really interested in this topic because they are not yet experts on it. In this chapter, you will see a lot of different perspectives on relationships and how to make them work. Our points of view have changed in many different ways. I'm sure yours will too!

Lilly Santana

By talking to each other.

How can a teen relationship work? Relationships are something that almost every teen goes through. I interviewed my friend, Susana, who had a few things to say. Susana has been in and out of relationships and she knows a lot about this topic. Something she told me was, "Try hard not to argue," and talk about each other's differences before they lead to bigger problems.

Susana used to go out with a guy named Jacob. In the beginning, Jacob was a really nice guy and used to take her out everywhere and buy her lots of things. After dating awhile, Susana started to notice that Jacob had a real ugly side. "He used to scream for no reason," she said. He always wanted everything his way and always wanted to be right. One time, he wanted Susana to cut school. He wanted her to

SUGGESTED RESOURCES

Books:

Chicken Noodle Soup for the Teenage Soul by Jack Canfield

Magazines:

Seventeen

Teen Vogue

Movies:

A Walk to Remember

Aquamarine

Boys and Girls

Real Life Teens: Lessons of Love

go his house instead, but Susana didn't want to. Because she did not do as he asked, he dumped her. She told him that they could try to work this out but he replied, "No." It was clear that he only wanted things his way and did not care about Susana's feelings. The fact that Susana wanted to try to work things out is an example of how to make a teen relationship work.

I, myself, have been in relationships that haven't turned out to be so good. I think that the best way to work out a teen relationship is by talking about each other's problems. Another way is by taking things a little more slowly. Often girls are the ones who get hurt in a relationship because some boys just want sex and nothing else. A lot of parents don't approve of teen relationships because they are afraid of their kids having sex and ending up with a child. I think that it is really important for teens to talk about these things because we all go through these situations in our relationships and it helps to hear what others have to say. It's better to learn from these situations than to end up getting hurt.

There was a guy I used to go out with. He used to argue about every little thing. He did this on purpose and it really bothered me. One time, we had a really big argument because he asked me if I liked his outfit and I said, "No." I really can't have a boyfriend who would be so insecure and argue over silly things. I broke up with him and told him that I was too young to be dealing with these types of problems. After all of that, it turned out to be a good decision to leave him. If you are in a relationship and you see that your partner is not willing to agree with you on anything, you might want to end the relationship right there before things get worse.

Susana learned her lesson and we all can too. She knows many ways to make a relationship work. But, she says, "I can't do it by myself." Both partners should give 100 percent.

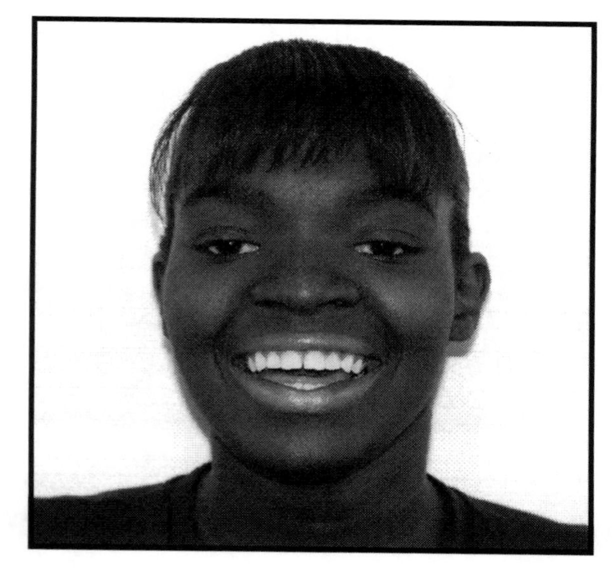

Faith White

By staying the way you are.

I have been a teen for a while now and have been in teen relationships before. When I was in one of my relationships, it made me do things and act like I was on top of the world. I did not really care about anyone unless it had something to do with me and I was getting something out of it. I was just 15 years old. Thinking back on my past, I am really happy with myself.

I decided to interview an adult to see if she acted the same way I did as a teen. I met with Ms. Williamson, a guidance counselor at my school, to get her story.

When I asked, "Why do teens want to be in relationships so much?" Ms. Williamson replied, "They sometimes just want to talk to someone else besides their family." She looked at her desk like something was wrong with the question, so I asked her if something was wrong. She answered, "Nothing's wrong. Just some old time memories came from the past."

Next, Ms. Williamson looked at me with her eyes open really wide. She offered me a little advice. "A male ego depends on its money. If a guy is going out with a girl and doesn't want his girl to make more money than he does, that is what they call pride. Guys like to be able to say, 'I take care of my girl.' When boys or men start saying things like that, then that's when their ego and pride comes in." Ms. Williamson made me think more about that. It made me realize that I have been in that situation before.

We came to the end of our interview and both wondered why it is that teens and adults look at relationships so differently. Ms. Williamson and I decided that it just depends because everyone has his or her own opinion. Everyone does not think the same. People are different and that is a good thing. If you are a teen and you are reading this, I want you to know something teen-to-teen. Stay the way you are. Do not change for anyone—no one but yourself—and that's real.

SUGGESTED RESOURCES

TV shows:

Boy Meets World

Sister, Sister

Music:

"Leave" by JoJo

"Sex" by Lyle Jennings

Websites:

http://www.arts.yorku.ca/lamarsh/projects/trp/index.html

http://www.lfcc.on.ca/teendate.htm

Discussion Questions:

1. When do you think teens should get out of a relationship?
2. Is there a good way to break up?
3. What role do family traditions play in the life of a teen?
4. What should you do if you see a teen getting too caught up in a relationship?
5. How can you comfort a friend who is suffering after a breakup?

How do you resolve conflicts with peers?

This chapter looks at how teens deal with conflicts. From teens to teachers of all different ages and with different perspectives, we talked to people about how school conflicts can start as well as what can happen during a conflict and how it might be avoided. All teens can stop a conflict, but some don't want to while others try to help, sometimes by bringing in an outside third party or an adult. We write about how we see this in our eyes because we know; we see conflicts almost every other day. Some turn into violence, and some are solved with communication. Which do you choose?

Darlina Arias

By ignoring personal attacks.

Conflicts are very common among peers, especially at school. In my experience, they often involve girls. The roots of the disagreements are often tiny misunderstandings that lead to huge blow-ups. I, myself, have witnessed many conflicts in my school, which intrigue my curiosity more and more each day. I decided to research about this topic and share my findings.

How would you feel if you were sitting in a classroom, minding your own business, when suddenly someone smacked you across the head? My best friend was sitting in her classroom when one of our old classmates went in and pushed her against the desk and punched her in the head. My friend fought back; she didn't know why someone would hit her out of the blue.

According to some research I have done

through networking and speaking to my peers, concern for this type of arbitrary violence in school is increasing. Classroom teachers and school staff often spend too much time and energy trying to deal with these classroom conflicts, and often do it in inappropriate ways. In response to this, conflict resolution and peer mediation training programs have also been growing. Programs like these have been organized by researchers of conflict resolution, non-violence spokesmen, and members of the legal profession because they have seen that teens can't always solve conflicts on their own. They need help.

"The issues of fighting, violence, and gangs have been a major issue in local public schools," says Elam, Rose & Gallop, peer mediators who have taken part in the National Association for Mediation in Education. There are negative outcomes to poorly managed conflicts, which include harmful effects such as stress and challenges to self-esteem. Also, according to the National Association for Mediation in Education, there were approximately 2,000 conflict resolution programs in United States schools in 1992, and 5,000 to 8,000 in 1994. However, according to the research I did, those programs are not always successful.

"Anywhere there are people, you're going to have conflicts. School is a society with people and when people's needs are not met, conflicts arise," said Ms. Williams, a peer mediator and teacher at our high school. Ms. Williams is considered by many to be an outstanding peer mediator and mediator trainer. "You cannot prevent all school conflicts. What we try to do is help resolve them. We mediate between disputants, we identify the conflict, help the disputants come to an agreement, and hold them responsible for that agreement," she says. Ms. Williams has been a peer mediator and a mediator trainer for many years and has mediated many conflicts. "I'm one person and I won't give advice to another person who is totally different from me. I have different views and opinions but I help people realize what they have in common and come to an agreement." I agree with Ms. Williams when she says that many conflicts can't be resolved because of the attitudes people choose to take. If the person chooses to be mad and stay mad, the issue cannot be easily resolved.

In schools, conflicts cannot be prevented because there is always competition, battles between power, discipline issues, and problems with morale and in students' personal lives. All of these play a role in conflicts. When I think about all the conflicts I've encountered, I realize situations could have been better if I had made different choices. In my freshman year of high school, I had an argument with one of my classmates after she made remarks that I didn't appreciate. When she met up with me after school, I tried to avoid a fight by talking her out of it. When that didn't work, I tried to go around her but her mind was made up. Her decision was to fight me, so she swung and I had to defend myself. It wasn't clear who won at the end, but it doesn't really matter. Violence, in my opinion, is not always the answer. It can

SUGGESTED RESOURCES

Books:

To Kill a Mockingbird by Harper Lee

The Chocolate War by Robert Cormier

The Outsiders by S. E. Hinton

Speak by Laurie Halse Anderson

Seeds of Peace by Sulak Sivaraksa

Magazines:

Seventeen

Movies:

Freedom Writers

The Outsiders

Romeo & Juliet directed by Baz Luhrmann

TV shows:

Saved by the Bell

Degrassi

That's So Raven

Zoey101

Local news reports

Internet:

http://peacecenter.berkeley.edu/Fall04_PeaceGames.pdf

Games:

Gaming the School System

Class of 3000

just make matters worse. Looking back, if I would have ignored my classmate's remarks, I believe that our confrontation could have been avoided. I would tell other students that if they don't want to have a conflict, don't listen to other people's comments. If you know it's not true, then don't pay attention to it. The one who wants to fight is the true coward, not the one who walks away. That person is the smart one.

Stephanie Rodriguez

Is violence the way in?

I'm interested in the issue of school conflicts because it seems the fighting rate of all teens in schools is going up. School conflicts are an everyday thing in public schools. I decided to go straight to the source and interview two teens with totally different perspectives. Benjamin, 17, and Christine, 16, are both high school students, but they have different views about how a fight is brought up and handled. Both have great personalities and loving parents, but were brought up differently. People say you act the way your parents raised you to act. Sometimes this is true and sometimes it is not. A boy's mom can be a drug addict, and he can still get good grades, still have a life, and move on to greater things. For example, Benjamin's mom used to sell drugs right in front of her three children. She thought she was doing what was best by getting money, but she never opened her eyes to the damage she could do. Benji stood tall and never let it bother him. Today he is becoming a great man; he is ahead in school and looking forward to a bright future. Benji is living proof that you do not have to follow in your parents' footsteps. Even if your parents do not set a good example, you can make something out of your life.

Though he is successfully overcoming his hard childhood, Benji is a fighter who thinks arguing and fighting is solving. He's the type of person who doesn't care how people see him. He needs to prove his point of view by fighting. Christine, my other interviewee, has a different perspective. She says fighting can be avoided. She believes that you can always ask for help from someone more mature and experienced. Christine says fighting gets you a messed up face, a record, or a suspension, and it doesn't help you succeed in life. She's never had a physical fight, and intends on keeping it that way, because she sticks to herself and doesn't pick fights.

"You a p---- a-- n-----!" This sentence of disrespect is what started Benji's fight, a boys' fight about who has more power. After this was said, Benji stepped back, turned around, and started throwing fists. People crowded around and started screaming, "Fight!" People Benji didn't even know were there, watching and enjoying everything. Punches to the face, bodies thrown against the gate—it all went down. In Benji's perspective, the fight wasn't ending anytime soon, nobody was winning. He said the fight started off two years ago because of simple looks from guy to guy and then some gossip. Benji can't stand fake people who lie, especially when they talk behind his back. He said the fight went down quick. Security came. Someone ratted them out and they both got a long suspension. Benji didn't mind because he believed "the kid got what he deserved."

Surprisingly, the kid and Benji are cool today. They see each other, and casually say "what up" to each other. Now they play basketball, the sport that brought them together. He has no choice but to forget the past and move on unless the kid wants to keep on. I asked Benji if he regretted fighting and he said no, but he regrets listening to instigators.

After hearing both sides, I think I've learned that although fighting leads to nothing, walking away looks bad, so always rely on the people closest to you and compromise. Violence really isn't the answer.

Demetria Thomas

Try to find a resolution.

All my life I had to deal with school conflicts. I will never forget moving into a new neighborhood in Brooklyn. I didn't get along with anyone. I went to a school with kids that really didn't know how to act and were rowdy and very tough. Being in school conflicts affects your everyday school life. It affects your grades and your behavior.

The person that I interviewed is a person who deals with school conflicts everyday, a high school guidance counselor at my school, Institute for Media and Writing. He believes that conflicts are a part of life: "We don't need to avoid conflicts; we just need to learn how to deal with them in a responsible way." One of the many incidents the guidance counselor had to deal with was a conflict between two students. He mediated by speaking to them individually, and it turned out it was a misunderstanding. In most cases that is why people get into conflicts. "Hormones, physical proximity, and being around each other together all day long leads to tension, and teens don't have much experience controlling their temper. That can lead to lashing out into a situation that they are upset about," the guidance counselor says.

"In life we can make mistakes, but if we can learn from them and not repeat them we can grow," he adds. "A lot of learning happens during conflicts. It's an opportunity to see what went wrong and how conflicts can be handled better." I'm glad I had the time to interview this guidance counselor because we have the same perspective on school conflicts. I believe we think the same way. We both believe every day is a learning process. From my own experience with fighting, the outcome was never good. Like the guidance counselor said, most fights arise from a misunderstanding. Fights can be resolved without getting physical unless you are pushed into it, like if the other person throws a hit. After getting hit, people feel forced to fight back. But there are a lot of other ways to handle school conflicts.

It's best not to fight in school. It only leads to more problems when you have a fight, win or lose. I hate looking back on school conflicts or any conflict I have had and regretting the actions I've made. I can't be the one to judge the position another person is in. I suggest that everyone involved in a conflict should do the right thing for their particular situation. But, remember everything you do has a consequence!

Discussion Questions:

1. If someone were to try to fight you, how would you handle it?
2. Do you think walking away from a conflict or fight is the right decision?
3. Would you be friends later on with a person you had fought with?
4. What type of fight do you think is justifiable, if any? Is fighting compatible with your beliefs?
5. How would you improve the way school conflicts are handled in your school?

Chapter 3

School Days

What factors prevent students from learning?

Sometimes teens think they know everything and don't need any help. Often, when someone offers a teenager help, they turn him or her down, saying, "No, I don't need your help. I am fine." But teens need to learn to take advice, especially if that advice will help them in school. There are many things that can disrupt our learning and here is some advice from teens, for teens, about how to stop that from happening.

Aaron Garcia

No food! No work!

How can students go through their day without food or something to drink? Just think about it. So many students do it, but how can you concentrate on learning and preparing for your future if you are distracted by thirst and hunger? The answer is simple: you can't. I know I hate going through the day without eating. I eat breakfast every morning and it enables me to focus in class. I start to get a headache and I don't want to work if I haven't eaten. I don't eat the nasty school lunch because then my stomach starts to hurt, so I'm stuck with an empty stomach until I get out of school. Waiting until 3:00 means I go seven hours without food! Do you know how hard it is to go through the day without having eaten anything? It's very hard!

I checked out an article from Cornell University to find out how not eating affects learning. The scientists found the following and it scares me:

Hunger and poverty in the United States are severe enough to significantly impair the academic and psychosocial development of school-age children and adolescents. Children from food-insecure families were found to be five times more likely to attempt suicide; four times more likely to suffer from chronic, low-grade depression (dysthymia), which is a high-risk factor for major depression; were almost twice as likely to have been suspended from school; and were 1.4 times more likely to repeat a grade and to have significantly lower math scores.

I don't need an article to tell me it's hard to make it through the day without food, and I'm not the only one that thinks so. I interviewed Sonia Mason, a nurse who has two daughters in college. I asked her what the effects were of not eating throughout the day and she said, "It's not good to not eat. It is hard to concentrate in school and you can't do your work." I asked her what you should eat for breakfast to get through your day. She said, "The best way is by eating a healthy breakfast consisting of eggs, toast, milk, and plenty of protein." Protein makes you feel fuller longer and it helps build muscle. If you just have bread and sugar, your energy drops by mid-morning. Some other important foods that will help you learn more include fruit, like oranges, and vegetables, like corn.

I asked her if she thinks school lunch is healthy and she said, "Not all the time because most of it is frozen or it has a lot of fat." If school lunches are not nutritious, it will affect the health of the students, which could result in their becoming ill more often or overweight. I went on to get her opinion on the taste of school lunch. "Most of it is nasty," she said. "But the pizzas are pretty good." Too bad tasty pizza is not necessarily nutritious food. It's just like if you eat McDonald's everyday. I know I'm probably the number one fan of McDonald's, but it's not good to eat fast food everyday, no matter how good it tastes. If you have too much fast food it can clog up your arteries. You can have very bad blood pressure and a lot of other bad things can happen to you. How can you learn if you're in a hospital or dead? I'm not saying not to eat it at all, but it should be viewed as a treat you have once in awhile.

Ms. Mason makes some suggestions for what to do if your school lunch is not very edible. "When my kids were going to school, I always packed them lunch so they didn't have to eat the school lunch." If you pack your lunch, try to include protein, such as chicken, or a sandwich and some juice or water so your brain is ready to work in the afternoon.

Ms. Mason feels the same way I do about the importance of eating throughout the day. I know some people don't think food is an important issue, but it's actually an extremely important subject because it directly impacts our education. Bad food habits are responsible for your headaches, your sleepiness, your stomachaches, and many other things. Simply, we need to eat well in order to succeed in school.

Source

Lang, Susan. "Not always having enough to eat can impair reading and math development in children, Cornell study confirms." Cornell University: Chronicle Online. 22 Dec 2005. http://www.news.cornell.edu/stories/Dec05/food.insecure.learn.ssl.html

SUGGESTED RESOURCES

Books:

Tears of a Tiger by Sharon M. Draper

Movies:

Remember the Titans
The Breakfast Club (R)
Freedom Writers
Dangerous Minds (R)
Coach Carter
Mean Girls
Take The Lead
Half Nelson (R)
Class Act

TV shows:

Fresh Prince of Bel-Air
Smart Guy
Boy Meets World
Saved By The Bell

Music:

"They Don't Care About Us" by Michael Jackson

Julia Marroquin

Certain relationships can hold you back.

Most of the time, being in a relationship when you're a teen is hard because parents are often worried that the distraction of a relationship will prevent their son or daughter from getting a good education. My friend, Melissa, has a few things to say about this. Her parents are divorced, which affects the issue even more. She lives with her mother, who is like the "witch" in one of those fairytales. To make matters worse, Melissa is forced to take care of her two little sisters. You'll be surprised to see how this 15-year-old's family life has prevented her from learning.

I feel that being too closed in makes us want to break out even more. Melissa is never really let outside, except if she's looking after her two sisters. She's not even allowed to wear fun clothes. She often says to her mother, "Can I please wear something different today besides baggy clothes?" but her mother will often choose her outfits for her so that no one will be attracted her. If Melissa sneaks nicer clothes or has her nails too long, her mother will beat her. Her father does not live with her, but she likes to sneak visits to see him because he gives her nice clothes. She says, "I'm lucky my mom still hasn't caught me!"

At this point in her short life, Melissa only really had her friends and the boyfriend she had just started seeing to depend on. But once again, her mother being the "witch of the fairy tale" forced the parents of her friends to stop their children from talking to Melissa "or else." Even though Melissa's friends were told to stop talking to her, they weren't completely gone; they just couldn't really see each other like before. And since her mother didn't know about her boyfriend, she couldn't say anything. Melissa and her boyfriend were fine until they got deeper into their relationship and he wanted to see her more. This wasn't going to happen since Melissa's mother would flip out. He kind of understood this but still wanted to see her more often but since that wasn't happening they started falling apart and Melissa refused to let him go. The only way to see him was during school hours.

This was Melissa's first real relationship and she felt she had to keep him by her side. School to her wasn't that big of an issue anymore. She would miss school constantly to go see him and, when the report card came, she would run to the mail and stuff it in her shirt. She risked a lot doing this but it wasn't enough for him. He wanted more! Melissa couldn't think of anything else but to please him by having sex with him. "I know I did the wrong thing but I felt us falling apart and that couldn't happen…I didn't really have anyone else…I needed him. He would always listen to my problems." But after all the hard work of sneaking around and trying to develop a relationship, her world fell on her and she was caught by the witch of this fairy tale!

Personally, in my own situation, I actually did think being with my boyfriend was worth missing school. When I started spending all my time with him, it was great but when it was all over, I had nothing to look forward to. Like that saying, "Boys come and go but friends stay forever"…well, I am rephrasing it: "Boys do come and go but even though it's never too late to bring yourself up, you wouldn't want to be doing it while everyone else your age is working jobs while you're still struggling to get through college." And yes, I do believe that parents or guardians should never close their teens in too much because even if they say no, some teens will end up doing it anyway. Parents should try and keep an open mind. And maybe just sometimes instead of being a parent, try and be a friend that your teen can rely on.

Jose Ruben

Schools, teachers and students need to focus on the task at hand – education!

There are many factors that prevent students from learning. I have lots of experience with feeling pressured because I try to do my best work on time, but it's not always easy. That's why sometimes we don't put our best effort into our class work because it's too much. I, myself, have my own frustrations that prevent me from learning easily. At the end of each school day, when my father asks me, "What did you learn today?" I often don't have much to say. I have an especially difficult time focusing my attention on tasks that require math skills. I also find it hard to write down all the notes from what teachers are saying. There are days when I'm thinking of other things, like a story I'm developing, and it's very hard to focus on class work. This is something most teens would relate to. This is why schools need to take steps to make learning easier for students.

To find out specifics, I spoke to Dr. John Loonam, the director of our small learning community. As a former teacher and now principal, he has a lot of experience helping kids learn and so he knows what he's talking about. He is a good man and a nice person. His favorite subjects in school were social studies and English. He wasn't a quick learner, but he thought that mastering the subject was a better approach, and look where it got him today!

Speaking from the experience of having been a student, he believes that there are specific things that students do that keep them from learning. They often don't plan ahead and that can make it hard to get work done. Also, they don't take advantage of the extracurricular activities that can enhance learning in the classroom. He suggests that students join a club, like yearbook or student government.

As far as school administration, Dr. Loonam's perspective is that sometimes schools do things that keep teachers from teaching. As a director, he thinks that it's important to focus on supporting teachers in the classroom and "to avoid distracting them." This way they can really help students learn. He also wishes that there were more computers and other technology available for teachers to use in their classrooms. This could really make learning fun.

After talking to Dr. Loonam, I realized there are some things I could do to learn more effectively in school. For example, I need to listen to what my parents have told me about reviewing things I have learned so I'm more prepared for tests. I should also set aside a specific time each day to do my homework and revision. This, after awhile, will turn into a study habit. I also find that doing my own research on a particular lesson can make the lesson more interesting and will help me understand it better. For example, when I was learning about imperialism in history class, I did research on Japan because it always interested me. That made me understand imperialism better. As for math, I think it's time for a tutor!

Sometimes we students think we can handle everything on our own and stuff it into our heads at the last minute. But now I know I need to get help at the first moment I don't understand something. I think if I start taking these steps, I'll be on track to succeed as a student. Then I'll be able to reach even greater goals.

Marlon Thweatt

Distractions! Distractions! Distractions!

Ms. Batista taught 7th and 8th grade for five years. Last year, she was an English teacher and dean. Now she is a director of a small school. Ms. Batista has good ideas about how to teach teens. She believes that you have to make the work a challenge and also make it a fun process. Ms. Batista liked being a dean because she got to interact with more students and so she knows what may prevent them from learning in school. I know Ms. Batista as a loving person who cares about others but when I first met her she was a hard person to deal with and it killed me. I didn't like her because I thought she was out to get me when I got into trouble for threatening a teacher. The other dean was going to suspend me for 50 days but Ms. Batista said, "Don't listen to him, just take a week and call it a day." And I was like, "Whatever." And from then on, I thought of Ms. Batista as the kindest person I've ever known.

Having experienced working with teens for many years, she says, "Each student is loveable and students want to learn." But schools need to be an exciting place for students and they need happy teachers. Ms. Batista knows the power of a good teacher. The person I interviewed taught me things about life. Ms. Batista touched my heart when she said, "Each student wants to learn" and that touched me so much because there are some teachers who just want to get paid. Not all the teachers just want to get paid. They also want to help students get an education and a better life.

"There was a student whose English was not that good and failed the Regents twice. Once I cut time out to help this student, he passed the next Regents with flying colors." She is very aware of teen issues. She says, "Kids can be distracted by peer pressure, with getting hot new gear, and then they forget about the bigger prize—education!" Ms. Batista sums it up well, "Knowledge is power. Power is knowledge."

School issues are not really a problem for me, but other people in school do make it harder to learn. Some students start the problems with other students, by fighting, cursing, and threatening each other. Some come to school with family issues and take it out on the students and teachers. Finally, other students are bored because they don't feel like they can succeed. They mess with the students who are working and distract them from learning. That same student who was distracted becomes the problem for the teacher and the whole school. That's when the good students change and become more of a problem. Sometimes teachers, though not all, become lazy and don't want to teach the students. On the other hand, students sometimes don't let the teachers teach and that's hard on all of us. Students can be like predators, waiting and watching.

I have learned through my own experiences, and from what Ms. Batista has taught me, that people should make changes in their own lives and in the schools themselves to make learning easier and more fun.

Discussion Questions:

1. What do you think prevents students from learning?
2. How would you solve these problems if you were in the same situation?
3. What do you do when your own learning is disturbed to get back on task?
4. What advice would you have for parents and teachers whose children and students aren't learning?

When are school rules fair?

As young kids in high school, we think that we are all grown up and that we are free to do whatever we want. This is not always the case, and this is when school rules come into play. We tend to get mad when a dean or a teacher tells us to take off our hats or not to chew gum, but these things are all a part of following the rules, which can lead you to success in the future.

Danny Elmsox

When they reflect the goals of the school.

I have broken a few school rules, but the one I remember most occurred in eighth grade when I left my classroom without permission. I guess I sort of snuck out because I had a family situation and my brother had to pick me up for an emergency. Unfortunately, I didn't tell anyone and, as punishment, I got a phone call at home. Now I see that following the rules is very important.

The person in our high school who knows the most about school rules is Mr. John Angelet, our principal. I spoke to him about the subject and he was eager to fill me in. "First of all, I love this job and I intended to have it since I was a child." This gave him plenty of time to think about the importance of rules in school. Even when he was younger, he says that he followed most of the rules, but he was also just like any other student.

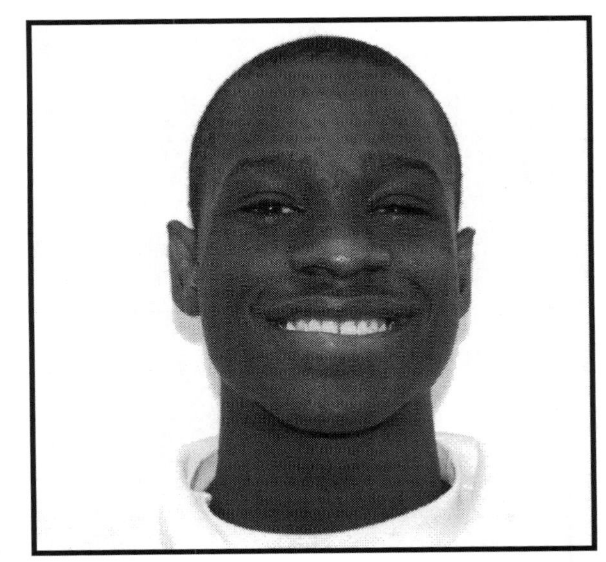

"But when are they fair?" I asked him. In Mr. Angelet's perspective, school rules are fair when they are enforced with the students' best interests in mind. He explained that it is his duty to make sure every student prepares for and has the opportunity to attend college. To do this, students need to be able to follow rules. To make rules fair, school leaders need to think them through well and choose them wisely for specific outcomes. Then they need to be enforced fairly. Most school rules come from the New York City Department of Education conduct book that all public schools follow. All of them, according to Mr. Angelet, have a specific purpose—to keep order and safety in classes so students can learn. He feels that without the rules there would be no organization. I asked him what would happen if a student constantly broke the rules. "They would be punished with a phone call home," he said. Other consequences include detention, suspension, and, rarely, transferring the student to another school.

Mr. Angelet's perspective on when school rules are fair makes sense because, if rules are not enforced, no one will follow them. His perspective on school rules reinforces my own feelings. The reason why I think many students do not agree with school rules is because sometimes they can be annoying, especially the little rules like no hats and no chewing gum. I personally feel that these rules should not be emphasized as much, but if a student breaks a big rule they should be punished. I feel that Mr. Angelet understands what the students need in terms of rules and punishments because they sounded pretty fair when I was interviewing him.

When I left that class early and got the phone call at home, my parents and I had a long conversation about how following the rules is important. I soaked in the conversation and applied it to school. Nowadays I try not to break rules, not even the small ones, because I want to mature and progress and eventually have a great life. Following school rules can help me achieve that.

Robert Vargas

Always, because they lead you in the right direction.

When should we obey school rules? As you get older, the rules change overtime. When I was a young boy in elementary school, I got into many fights. As I got older—when I was in middle school at Nativity Mission School—it was mandatory to wear uniforms and they showed me how to be a leader, and that's when I understood that rules are important. Now, as a public high school student, the rules seem more relaxed and the enforcers of rules are often laid back. It's such a significant change from my younger years. So my experience following rules has been like a roller coaster. I tend to follow what my friends do because friends have such an effect on my life. I just hope I don't affect my friends negatively the way my older friends affected me.

Anthony Davila is a soldier in the military, a CSI Detective in Bridgeport, Connecticut, a father of two, and my brother. He has his own perspective on our group's question and how school rules affected him while he was in school. "School is supposed to be a job to prepare you for the next level," said Anthony, when I spoke to him on the phone. He went on to explain that, "It's the kids' choice to be productive or not" and follow the rules so they can get to that next level. He also stated that schools should be stricter and more disciplined. Many students in my school and your school should understand the consequences of disrespecting teachers and having poor manners in a place you are attending for the entire day. This is why Anthony said, "Kids in school should where uniforms because, for example, students who wear fly clothes may get jumped." This can lead to conflicts in the classroom and prevent learning.

SUGGESTED RESOURCES

Movies:

Ferris Bueller's Day Off

Dangerous Minds (R)

Stand and Deliver

TV shows:

The Wire

Music:

"I Know I Can Be" by Nas

"Being or living in the right environment may show who you really are around your peers. As you get older, align yourself with the right people." He even stated, "The people you hang out with affect your life." As a young adolescent he saw many of his friends provoke fights and saw them get arrested. He segregated himself from them and explained, "My friends were lacking a sense of conscience. Those kids that I called friends constantly cut class and disrespected teachers. Now they are dropouts and don't have a life. What I'm trying to say is that cutting and breaking rules will get you in a lot of trouble with the law and your family. I just hope kids understand that there is a point to having school rules." They help you become more mature and they make sure that learning can go on.

As the years went on, he became an honor student and he signed himself up for the navy without his mom knowing. He did this because, he explained, "I wanted to change my life and start something new." The discipline of the military led him to join the Bridgeport police department and, specifically, the DARE program that visits schools and warns against using drugs. He knows the consequences of breaking school rules and doesn't want students to learn from experience.

I thought about cutting class and playing hooky many times. The day I interviewed my brother, I realized I should change my perspective and actually be somebody. His ability to follow the rules helped him achieve his goals. School is the place to develop that discipline because the education you receive there will get you somewhere in life. If teachers are going to be there for me I have to take advantage of that. "Teachers are not paid enough and they should get paid more for the time they take out to teach students," Anthony said. That's reason enough to follow the rules.

Discussion Questions:

1. Why do you think we should follow school rules?
2. Should school rules be followed off of school grounds?
3. Describe a time when you disobeyed a school rule. Why did you choose to do this? What happened?
4. Which school rules are intended to help us learn?
5. Which rules in your school would you most like to change? Which rules would you put in place, if you could?

Chapter 4

Thinking Ahead

Do you have to go to college to be successful?

College has always been a big issue for kids. Most of all our lives, we've been going to school and preparing for college. Ever since we were little kids our parents and teachers have been stressing how important it is for us to go to college, though a lot of people aren't sure if college is for them. Some people believe that they don't have to attend college in order to get a good job, but some people think that they do. Everyone has different opinions on the topic.

Rakiaunique Bazemore

My mama says you do!

My mother, Stacey Smith, grew up in the tough neighborhood of Harlem, New York. She earned her high school diploma, a Bachelor's degree in Human Services, and two Masters' degrees in Education and Special Education. My mom says going to college changed her life for the better and it opened up many doors because people took her more seriously. For example, with all her education, she was able to find a high paying job that she likes. My mom now works in a daycare program with disabled children that need extra care. She was able to move out of the tough neighborhood that she grew up in and move into a house, where she was able to make a better life for her children. Our family doesn't really look at her differently. They are happy for her. My grandma is happy to see that one of her children is successful. My mom says that growing up in a tough neighborhood helped her choose the field she wanted to study in college, which is special education. She thinks that if she hadn't attended college she would probably be a low-paid cashier. My mom says, "If you want to make a better life for yourself you should really attend college. It's a great

experience and it opens many doors." I'm proud to have a mom like her.

Not all college experiences are as positive. I have seen people who completed high school, received a degree in an excellent field, and still aren't doing anything with their lives. My question for these people is: Why put yourself through all the hard work in school and then not use your knowledge to empower yourself and others? College is expensive and you need to be successful to make the investment worthwhile. If you don't, you'll end up wasting all the money you spent.

One of my family members was so happy about going to college. He did everything he had to do; he graduated from high school and received a Master's degree in Writing and Journalism. My family pressured him into this situation and made him feel like he had to succeed. He did what he had to do, but didn't do it for himself. He did it for everyone else. Now he just sits at home all day watching TV, living off government programs, and getting money from the state. My family is very disappointed in him, but he just says, "I did what you wanted me to do and now I'm doing what I want to do." Many people feel they need to finish school and go to college to make their families proud. They feel lots of pressure to be perfect. I say you should follow your own dreams, not someone else's dreams.

As far as my mother goes, she's also now enrolled in a culinary school. My mom believes you can be successful without attending college, but she would still suggest college to everyone because it's a great experience. You learn more and you meet new people. I think that the harder you work in life, the more you'll receive! My mom would agree, "College is the way to a better life and future."

Having a mom like this affects the way I think. If she could do it, then I can do it. Her influence makes me want a better life for myself.

Dalila Fonteno

No, but you should if you have the opportunity.

The first time I considered the question, "Do you have to go to college to be successful?" I asked myself, "How should I know? I'm not in college!" But this question was a concern of mine. I wondered sometimes, if I don't go to college, can I get a good job? I also thought of my grandmother. She was young at the time in history when African Americans didn't have rights and didn't have the same privileges that I now have. Every time she asks me, "Are you making your grade?" I tell her yes and she smiles.

Since my grandma didn't have the same opportunities as I do, I decided to ask her opinion on the connection between college and success. I felt her wisdom would help me face any challenges in college and remind me that I am going there to make a difference and to make her proud. My grandma talked about how she only had a seventh-grade education and never expected to go to college because most African Americans of her generation couldn't go. At an early age, she decided to move from South Carolina, where she was growing up, because it was not for her. She headed for New York City because she felt, at the time, it would be a better place for her children and, hopefully, grandchildren. My grandmother didn't want to stay with her eight siblings and have the same life; she wanted to be different from them. She wanted to prove to her family she would do fine in the city.

When she arrived in the city, she had no money and she had to stay with a friend, so education wasn't really on her mind. The only thing she really cared about was making a living and supporting herself. She said, "I didn't want to go to college because I would have to start high school. Who wants to be in school that long with a young child?" Not having the opportunity to go has given her a strong opinion on the subject of college. She believes there are people today who have the chance to go to college and they are not using it. The kids who can go to college don't really take the real opportunities they can have. My grandmother tells me how she doesn't know why kids today waste their lives in the streets instead of trying to become the first black president. There are so many ways to go to college. You can go to community college and you can still be successful.

I asked my grandmother how she defines success. "To finish school and get a good job that pays well," she answered. She thinks that if you go to college you should get as many degrees in college as possible. I asked her why she thinks that degrees will lead to success. She said, "It will help put money in your bank account and you can get a very good job, depending on the degree you get."

After I interviewed my grandmother, I understood how important college is. I listened to how she had to struggle throughout life and how it would've been easier for her if she had had an education. I think that if you have a roof over your head and a well paying job, you have reached a certain success, but I also think that I have to take advantage of college and get my education because my grandmother didn't have a chance to. When I finish college I will be the third person to graduate in my family. I want to prove the statistics wrong. I know I will make it and I will be successful.

SUGGESTED RESOURCES

Books:

300 Jobs Without a Four-Year Degree by J. Michael Farr and Laurence Shatkin

200 Best Jobs for College Graduates by J. Michael Farr and Laurence Shatkin

Movies:

Drumline

American Pie (R)

Accepted

Higher Learning

Legally Blonde

Revenge of the Nerds

With Honors

Back to School

TV shows:

A Different World

College Hill

Music:

"I Can" by Nas

"College" by Kanye West

MSU Fight Song

Maria Gomez

No, you can be successful without college if you're willing to work hard and challenge yourself.

"College isn't for everyone," says Alexa, an Hispanic woman who is making it in this cruel, tough, and, most of all, expensive world without a college degree. She was born and raised in midtown Manhattan and now works in the city. Alexa, only 29 years of age, is working at a major brokerage firm called Morgan Stanley and she earns more than many men and women who have spent two to four years at college getting degrees.

I've always believed that success depends not on the degrees that you have but on your will to succeed and how hard you are willing to work. Alexa agrees, "If you are driven to succeed in life, you're going to work as hard as the next person." Of course, certain professions depend on having a degree. There are no short cuts if you want to be a doctor, lawyer, or college professor, but success in many other fields is solely dependent on the work that you put in. Alexa speaks from experience when she tells me, "You just have to work hard and earn your success. I worked just as hard as a person with a degree. I have worked even harder than others." Alexa didn't start out working in the field she's in now. Previously, she worked for a car service and also for a few financial operations businesses, but the route she's taken to get there has shown her that success is dependent on your willingness to challenge yourself. It's about having faith in your abilities and being willing to learn. "At my first job, I had a nasty boss so I had to take a chance and quit. I still wasn't doing what I'm doing now. I didn't start until after a ton of interviews. Then I got a job that I've never done before. I told the man who interviewed me that I've never done this. I was honest because I didn't want to fail. So he gave me a chance. He told me he knew if he trained me, I would get it quickly. In other words, he saw a lot of potential in me. Appearance is everything. I understood what he taught me, practiced a lot and got better. Some call it luck. I call it hard work, but in order to work at

Morgan Stanley you must be eager to learn more everyday."

Alexa is in charge of processing corporate action events. Her willingness to learn from everyone around her and her hard work has ensured her success in the industry. She earns a great salary and is recognized by her colleagues as someone worth having around. I wanted to know what pushed her to do better and better. "When you work at a place surrounded by people with degrees, and you're the only one without one, you tend to feel you must prove yourself to everyone. You're eager to show that you're as good. Everyday's a challenge. It's a struggle to succeed, but you can always be someone important. The trick is to never to stop learning."

Latifah Lane

Yes! College is important! (If you're not relying on luck.)

I believe that college is very important, especially for a young minority, such as myself. Do you have to attend college in order to get a good job? I'm not very sure about that, although I do think that you have to go to college to get a job that you like, or to have a fulfilling career. I've always thought that if a person finds a good job without having a college education, it's by luck or chance. I believe that going to college also brings with it a lot of other opportunities.

My sister and I sat and spoke in the bedroom that she and I share, so I could find out more about her perspective on the issue of college. I really feel that my sister's thoughts on the subject are important because I have a lot of respect for her. She is an 18-year-old student in her first year at Monroe College. She is going to school because she wants to become a medical assistant.

My sister believes that college is important because "it helps you to further your education and helps you build a career." She went on to explain, "I do not believe that I would be very successful without a college education." I asked my sister how she defines success and she responded that success is achieving your goals and making a decent amount of money. So, based on her definition, college is an important step if you want to be successful. I asked her what type of job she thought she would have without going to college and she said, “I would probably have a regular minimum-wage job, maybe working at a clothing store or something.” My sister believes that by attending college she is gaining new experiences, meeting different kinds of people, learning to work well with others, and learning responsibility. “If you don’t go to college, you’ll miss out on a lot, including better job and work experiences.” She doesn’t believe that not having a lot of money is an excuse not to attend college. “They have so many different programs to help people,” she explained. “There’s financial aid and other types of student loans.” If you want to make it happen, it will happen.

I feel the same way as my sister and I agree with all of the things she said. I know other people who are struggling to find their way in life and really are trying to find out what they want to do with themselves. When I asked my sister what she would do if she didn’t have a chance to go to college, she answered, “Rob a bank!” I laughed, but I knew what she was trying to say. If she didn’t have this opportunity to get a college education, she would feel hopeless. I think that college is very important in order to be successful. It gives a lot of people the assurance that they can have the career or job they want.

Discussion Questions:

1. Who in your life has been successful without college?
2. Do you want to go to college?
3. Do you think you have to have a certain amount of money in order to go to college?
4. Do you think that a city or state college is just as good as a private college?
5. Do you feel pressured into going to college?

How do you choose the right college?

Before we started this project, we had four people who had different aspirations for the future. One writer's perspective on his future was turned around after writing his piece. Before, he didn't think that college was a priority,and now he's planning to pursue higher education. Another writer wants to skip higher education and go straight to work. Two of the writers felt that college was important and had plans to go right after high school. College may not be for some people, but for others it's mandatory. Different people have different ideas about how to go after their dreams.

Kathyrine Danyluk

Apply for colleges that meet every standard you have.

College education is viewed as a necessity for success in the future. There are around 25,000 high schools in the U.S. and it is competitive, but it is possible for most high school graduates to attend college. According to the National Association for College Admission Counseling (NACAC), "Nearly 65% of all high school graduates enroll in some form of secondary education within twelve months of graduation." Steven Roy Goodman, a top college and graduate school admission consultant, explains that statistics show this has increased from 1970 when it was 50% of high school students going to college. There are 3,000 colleges and universities in the U.S. With the pressure of getting into a college and so many options, all of these students face the same problem: How do they find a college that is right for them?

I'm still a sophomore in high school, so I don't have much experience with finding the right college yet. But I do have many concerns, so I've started looking for some information. I learned that when looking for a college you should make sure to check out the college office at your school. They will help you know what colleges are out there. It's also good to visit college campuses. In 2005, 74% of colleges hosted campus visits. This is one way to figure out if a college "fits" you.

"Apply for colleges that benefit every standard you have," says Cynthia Malagon, a student who got her college diploma (after four years of college) and is now pursuing graduate school (usually a two-year program) at Stony Brook University. In high school, her average was a 94. Her school was very competitive which made her grade stand out. "The college office at Midwood told me where to apply," she says. She applied to eight colleges so she would have a choice. When she finished applying she was very nervous. She was anxious to know which schools had accepted her.

When the letters started to come in, Cynthia had a ritual for how to open them. "I would first check the thickness and then I would have someone else open it for me." Her father implemented the idea of going to college. "He basically drilled it into our heads," she says. "I wanted a college that was not too far away from home, where I could visit but still dorm." Her father supported this. She also wanted a college that was known for a specific program and was affordable. Financial aid helped her pay a lot of the tuition. She also paid with whatever money she had left over from summer jobs. "Visit schools. Do not choose a school because it is known for parties," she stresses. "I have no regrets. I got a good education for a good price, met good people, got a great job and discovered I was made for business."

College can cost a child around $18,000 to $60,000 for four years, depending on whether they go to a public or private college. It's not easy to pay all of the tuition. People get financial aid for help (FAFSA is the federal program that loans money to students). According to the Washington Post, loans are now 70% of all financial aid packages. When choosing, keep the tuition in mind. A full 50% of those who drop out blame financial factors. If tuition is still a problem, it's good to try and apply for some of the 1.7 million scholarships or a college grant. There are $7 billion worth of college grants available.

As a sophomore, I still have one year to start looking into colleges. I started early though because I want to be prepared. I've looked into many scholarships and colleges and I plan to get financial aid, a scholarship, and a job to pay for college. I also want to study abroad because I have learned that you can get as good an education as you can get from the U.S. but for a much cheaper price.

From what I learned, I've decided that I have to start getting serious about my future. It's for the best. Researching the best ways to

SUGGESTED RESOURCES

Books:

America's Elite Colleges by Princeton Review

Colleges That Change Lives: 40 Schools You Should Know About Even If You're Not a Straight-A Student (Revised Edition) by Loren Pope

The Truth About Getting In: A Top College Advisor Tells You Everything You Need to Know by Katherine Cohen

Magazines:

College Board Review

Common Sense Guide to American Colleges

Insiders Guide to the Colleges

Movies:

With Honors

187 (R)

Perfect Score

Television:

Daria: Is it College Yet?

Saved By the Bell: The College Years

Websites:

www.collegeconfidential.com

www.collegeboard.com

www.collegeview.com

Games:

Kid's College Educational Software (U of M)

College-opoly

find the right college has helped me a lot. "About 54% of U.S. students entering college have a degree four years later," according to Goodman. I'm hoping this is what happens when I go to college. "When you are in college, make it the best years of your life, both academically and socially," says Cynthia Malagon. I hope that happens too.

Sources

Knight, Rebecca. "Ivy League Colleges Find 2006 is Buyer's Market". Top Colleges. April 14th 2006. Financial Times. 15 March 2007. http://topcolleges.com/news.html#ft

Goodman, Steven Roy "Hey, Profs, Come back to Earth". Top Colleges. April 10th 2005. The Washington Post, 15, March 2007 http://topcolleges.com/news.html#ft

Goodman, Steven Roy and King, Margaret J "PA's Big Test: Improve Dropout Rate." January 26, 2006, The Philadelphia Inquirer, 15 March 2007. http://topcolleges.com/news.html#ft

Careers and Colleges 2007 Alloy Education. 15 March 2007. http://www.careersandcolleges.com/

Priscilla Felix

Do your research.

College was something that had never been my first priority. As a matter of fact, it was something I never thought about. I always assumed that being a sophomore meant I wouldn't have to worry about it until I got to the twelfth grade. But I was completely wrong. My older sister inspired me to plan on attending college and to start searching for colleges now, instead of waiting until the last moment when it could be too late.

My sister graduated high school and started attending Marist College in Poughkeepsie, New York. When she first applied to this away-from-home college she hoped only for freedom, parties and guys. Like any other teenager, she thought being away from home was going to be the best. But after two weeks alone in a dorm with new people and everything she knew so far away, she became homesick. She would call home five times a day and she would say that she missed all her family and friends. After a few months of meeting new people and adapting to the lifestyle of being a college student away from home, she became comfortable. It turned out that her choice of college was the right one. Perhaps if she had chosen a school closer to home, things would have been different and her focus on school would not have been the same. She finally graduated with honors and is constantly telling me to start thinking about college so when the big time comes, I'll have something in mind.

This is where it all starts: "How do we know which college is right for us?" To find out, I interviewed Ms. Rodgers who is a parent whose son is now applying to colleges. Like any parent Ms. Rodgers wants the best for her son, so she's really involved in this big step her son is about to take.

I first asked her how she helped her son pick the right college. She told me that she did research, visited colleges, and attended open houses with him. She was on top of him to make sure that the colleges he was picking were an opportunity for him. She even said, with a smile on her face, that she probably helped too much. It is clear to me that parental involvement is important. She recommends that parents get involved by emphasizing to their kids the importance of college. They should also remind their kids that colleges see dean reports, tests scores, and grades. She also recommends that kids start thinking about college in the ninth grade, as opposed to the eleventh or twelfth. When I asked about money and loans and scholarships, she was really open. She said, "There are a lot of opportunities."

She told me that he's scared to be away from home. He thinks he might not be successful and that it might be too hard. He's tired of classes that he doesn't like. Ms. Rodgers's son is looking into attending schools in California, Florida, New Jersey, and New York. One of them he is interested in is Princeton University in New Jersey. That means that he'll be dorming on the campus, even though she doesn't think that it's a great idea because it will affect his grades negatively. It's not that she doesn't want to encourage him. She just doesn't want him to get his hopes up and get disappointed when things turn out to be more difficult than expected. To prepare for this, he also applied to other schools, which she thought were more suitable for his grades. She said that her child is mature enough to be on his own at college. When I asked about her college experience, she said, "I wish I had had a different college experience." She commented that she didn't attend college right after high school and she wished she had.

To tell you the truth, I had never thought of attending college until I interviewed Ms. Rodgers. She opened my mind to the idea that there are a lot of opportunities. Now that I know I'm going to college I have specific things I'm looking for. I want to get into forensic psychology, so I would like to attend a school that is respected in this area and offers science classes in the areas of psychology, human development, and criminology. These will give me the competitive skills I will need when I get a job. When searching for my college I would consider the programs being offered, the distance the school is from my home, and, most importantly, a college that fits well with my grades. If the school fits me, I know I will be able to meet all the challenges ahead.

Jofran Mendoza
It's about priorities.

"When you go to college, you're under a lot of pressure to graduate... or at least it was that way for me," said Jean, a student at Buffalo University, about how he felt when he was a freshman in college. "Since I was the first to actually graduate from high school, my parents were all proud of me and they expected me to go to college." Jean went on to explain how much pressure it was for him: "And then you have to get your master's or whatever it is you need. Anything else would be a complete failure." He said with his voice deepening with each word.

"I remember the day of my high school graduation and all the smiles that I put on my family's faces." Jean explained, "I was so proud of myself, but I knew this was just the first step. High school diplomas these days are worth nothing and my main goal wasn't some high school diploma. It was college, making something of myself so my parents wouldn't think their son was a nobody."

College has never crossed my mind before, but as I get older I'm noticing I'm not twelve anymore and I have to start thinking of my future because life won't wait for me. But what is the right choice for me? To be honest I don't even know what I want to be. I mean I keep changing my mind on what profession I should have. I never thought it was so hard to prepare just to go to college. On top of that, I don't know any other university other than Harvard and U.S.C. I was really stressed out when my mother lectured me about how important it is for me to graduate high school and go to college, but I personally didn't want to go. It's not that I don't care about my future, but it takes money to get into college and my mom doesn't have a lot of it. I'm not the brightest kid around, either. That's why college hasn't really been my first choice for my future.

But I have begun to think and I have been wondering what will become of me in ten years. I began to see what my cousin was talking about. My mom had given me everything I could ask for and in return she just asks me to study and become a professional who doesn't have to labor like she does now, trying to balance two jobs and keep up with her only son. I didn't know what to do. Since she's a single mother, I thought maybe I could get a job and chip in on the rent, but my mother is stubborn and she just wants me to focus on school.

And with money low, I wonder if I could even stay for one semester at a college. After doing a little research, I found out that, according to the U.S. Department of Labor's Bureau of Labor Statistics, between October 2004 and October 2005, 400,000 young people had dropped out of high school. 68.6% of high school graduates in 2005 enrolled in colleges or universities. I don't want to be a dropout and be some bum on the streets asking for change but what can I be? I guess I could become a writer. I like to write.

It seems that college is a lot more important than I had figured it to be. I mean, I always knew that if I go to college I would have a great life, but I was just too lazy to try in school. But now I think that I would like to be a writer and write a novel or two and become famous like Stephen King. First I have to do well in high school and, the biggest step of all, get into college and graduate. "College can be fun and seem like it would be easy, but those times come and leave quickly," Jean says. "The courses are hard and if you miss one class, it's like missing a whole semester. It's important to attend every single class and put aside your social life because, if you're serious about your life, you will never miss a class," he said with his eyes locked onto mine. "College is something you can't just give up on, if you truly want to be someone who people admire. To be a hero, you'll go to college." He took a slight pause and said, "If not, your only purpose in life is to die."

Sources

"College Enrollment And Work Activity of 2005 High School Graduates." Bureau of Labor Statistics. 24 March 2006. http://www.bls.gov/news.release/hsgec.nr0.htm.

Brittney Rodriguez

You first have to have the right mindset.

I wonder why people drop out of college. Why even start if you know you may not finish? When I asked my mother this question, she looked at me with regret in her eyes. She explained to me that "there are bumps on that road of success and some people are just stopping to fix their tires." That metaphor had me thinking the whole night. This generation has many pit stops. Lives need to be changed soon. My mind was just rambling about how many teens are struggling to get their education. We struggle through many pressures—family, health, and many more. We just can't stop thinking of these things. My mother was a part of this. When she was younger, her mother rarely let her go out. When she finally got her freedom, my mom just went crazy with the parties, the boys, and the friends. She really wasn't thinking of school when she got to college. Her mind was in another direction. The "wrong" choices prevented her from getting her degree. She wanted to become a nurse, but her dreams were not fulfilled.

I'm not here to tell you that you need to go to college or else. I'm just trying to tell you to keep your options open. I think we are scared to grow up with no car, no money, and no home. Some people actually do make it without college. But what happens to those who don't make it without an education and they don't know what to do? They have nothing to rely on and rarely will a company hire an uneducated person.

My mother's experience influenced me to stay in school so I could keep my options open. Some of those options would be the "wrong" ones and some are the "right" ones. The "right" ones are to stay in school and not worry about the boys or the parties. There is a time for all of that, eventually. Just not right now. Since I want to become a paramedic, I have to have lots of extensive training on top of a high school diploma. First, when I turn eighteen, I want to become an EMT. To do that, I have to take courses and obtain my EMT license. You don't need a college degree but you do need to finish high school. My mom's experience affected me by encouraging me to become more ambitious towards my goal of success. There are many opportunities out there for everyone. You just have to get off your lazy bums and go look. If you don't have the financial support, there are programs to help you. When I get bored of that job, I want to become a registered nurse, which requires a degree, which takes about two years to obtain. I want to better the world and have fun doing it. If you put your mind to it, anything is possible. You should go to college after you finish high school because doing it later makes it difficult to balance a family, a job, and education.

When my mother attended college, it wasn't meant for her because she was too immature to take school seriously. She also realized that high school didn't prepare her well for the demands of higher education, so she had to take some courses over. "I went to school and it was different. I became independent," says my mom. She had to care for herself, cook, clean her dorm, and make sure she got some studying into her busy schedule. She took a long leave of absence and decided she would go back later in life when she was more mature and settled.

Now my mother goes to community college and she says, "It's difficult to balance family, a job, and education." She really struggled to get where she is now, just studying to become a registered nurse. She has to take twice as many courses as she did before. Meanwhile, she works as a financial representative at Montefiore Medical Center. Despite having a very busy life, she's doing well. She just has to do what she has to do to get back on track. I, on the other hand, have a long way to go.

Based on my mother's experience I think that to choose a right college first you have to have the mind set for college. If you don't, you won't get anything done. From what my mother said to choose the right college you should consider the location, reputation, majors, cost, and schedule. My mother chose her school because she is able to attend classes at night and have a job during the day. She also didn't want to pay a lot. This school offers a good education at a relatively good price. She does wish, however, that she had chosen a school that is closer to home. The commute can make going to college difficult. If you have a job, consider night school. All in all, your choice of college will greatly impact your future so choose wisely!

Discussion Questions:

1. In your opinion, what kinds of aspirations do people going to college have?
2. How do family members' views on college affect your own view?
3. Is college necessary to succeed in life? Why or why not?
4. How important is college to you and why?
5. What are some of the reasons that people drop out of school? What could schools and society do in order to stop people from dropping out?

Chapter 5

Under the Influence

Are drugs and alcohol as dangerous as they are portrayed?

Weed, pot, Mary Jane, blunt, an L, kind bud, joints, trees, buddah, piff, haze, green, purple haze, the chronic, smoke, regs, brewsky, sauce, drink, drunk. These are all the names and terms that we know that refer to drugs and alcohol. You hear these words everywhere, every day. What you do not hear all the time is liver disease, cancer, alcoholism, incarceration, rehab, addiction, and memory loss. Teens have to be more aware of the dangers of drugs and alcohol by reading this chapter from front to back. In this chapter there are stories and facts, eye-opening statements and powerful perspectives that will change the way you think about drugs and alcohol.

Luisanny Almonte

Be aware! Drugs and alcohol can be harmful to your health.

My friends have had experience with drugs and alcohol that have not been very good. They have been drunk before and they went crazy. They did many things that they don't remember and said things that they regret saying and didn't mean. While they had alcohol in their systems, they felt great and happy and all they wanted to do was party. Unfortunately, they didn't feel as good the next day; they felt sick and had a headache. Their experiences with marijuana, the only drug they have tried, were not too good either. At the time, they felt cool, but afterwards they had a bad feeling. They didn't feel normal but sleepy, hungry, and too tired to do anything.

I interviewed a graduate student named Jennifer. This student's curiosity, like my friends', led her to try bad things. She has experimented with drugs and alcohol before. She believes that teens should be careful when they use drugs and alcohol because they are dangerous and addictive. Jennifer has used three types of drugs in her life: marijuana, magic mushrooms, and ecstasy. She believes most people drink and use drugs to have fun and to relieve stress. Not everybody thinks like she does and she is concerned that young people are not as wary as she is about the addictive substances. Instead of using drugs and alcohol to take her stress away, she will go for a walk and do other stuff instead. Jennifer stopped doing drugs because she knew the consequences and didn't want to be part of a life that was not going to get her anywhere.

Drugs and alcohol are not that easy to quit but it all depends on the person and their personality. Some people may stop when they want to stop. Other people may have an addictive personality so they find it extremely difficult to stop because they fall in love with the drug and feel that they always need it. People need to take caution with the kinds of drugs they take because each drug is very different and has its own effect. But all will hurt your health, relationships, wallet, and ability to succeed. Once addicted, you will need increasingly regular 'hits' because your body craves it. Drug addiction can change the direction of your life for the worse. As your craving to maintain the 'high' becomes the most important thing, it is likely that you will discard the goals and dreams that you once had.

Jennifer described for me some of the immediate effects of the drugs that she had tried. Marijuana makes you feel mellow, tired, hungry, and happy. Magic mushrooms makes you feel strange and makes you hallucinate, which can be creepy. Ecstasy makes you feel hot, sweaty, and also thirsty. It can also make you appear to feel really happy, but ecstasy is very dangerous for your brain if you do it too much. It's natural for teens to want to experiment but they should know for themselves the risks involved.

Drugs and alcohol are powerful and if you are going to use them then you should know what to expect. My friends' experiences using drugs and alcohol as teens were not good, and Jennifer as an adult also did not have such a good experience. The process of researching this important topic has influenced my own perspective. Now, I will think more carefully before experimenting with things that are not good for me, especially drugs and alcohol.

SUGGESTED RESOURCES

Books:

Random Family Love, Drugs, Trouble, and Coming of Age in the Bronx by Adrian Nicole LeBlanc

Bodega Dreams by Ernesto Quinonez

The Late Great Me by Sandra Scoppettone

The Glamorous Life by Nikki Turner

Breaking the Cycle by Zane

The Basketball Diaries by Jim Carroll

Go Ask Alice by anonymous

On the Rocks: Teens and Alcohol by David Aretha

Movies:

Paid In Full (R)

Thirteen (R)

The Basketball Diaries (R)

Menace II Society (R)

Reefer Madness

Requiem for a Dream (R)

TV shows:

The Wire, Season 1

New York Undercover

Music:

"Dope Boy Magic" by Yung Joc

"Push It" by Salt-N-Pepa

"Soul Survivor" by Young Jeezy

"Fury Unleashed" by Visceral Evisceration

"Lil' Boy Fresh" by Juelz Santana

Visitors:

Police officer

Drug and/or alcohol abuser

Nina Alonso

Drugs and alcohol may be as dangerous as they are portrayed, if you allow them to be.

Have you ever felt like you didn't want to go home? Living with a family member that does drugs or abuses alcohol can be rough for a teenager or an adult. It's hard to know exactly what the substance abuser is feeling or going through. Still, you can help the person be more aware of the damage their behavior is doing. Sit down with him or her and try to understand why they need drugs and alcohol so badly. Don't hate them because of their problem.

Living with drugs and alcohol abuse can hurt anyone. I spoke to a person who has seen it her entire life. Drug and alcohol abuse has had a great impact on her and the way she deals with her family. She has realized that having people in your life who abuse substances can hurt, and sometimes even ruin, your relationships with others around you. Her name is Elizabeth and she knows exactly how you feel.

Having her father, brothers, and sister abusing one thing or another has educated Elizabeth and has shown her the other side. She did turn to alcohol at one point herself. She was in a bad relationship with a man and she thought alcohol was all she had. Due to her insecurities, it was hard for her to come out of it herself. She turned her back on everyone, even her own family, so that nobody could judge her, or know what she was doing. She says she doesn't care for drugs because they destroy you and everything around you, but she has mixed feelings about alcohol. I, personally, think both can destroy you and everything around you because, when it comes down to it, nobody really wants to deal with you or your problems.

Feeling disgusted and ashamed, Elizabeth has separated herself from her family. Her brothers have been addicted to drugs for more than eight years. It's frustrating when you try to help your family and they don't want to accept your help. One of Elizabeth's brothers was incarcerated for almost eight years and was clean for seven of them. Coming back to his old neighborhood and his old friends turned him back into the person he was before he was clean. He showed the entire family he could not change and was unable to do anything good for himself. She feels that, ultimately, you can't prevent anyone from doing anything; no matter how much love and support you give them.

Speaking to Elizabeth has opened my eyes. There are some things people just hold back and hide about themselves. She doesn't let any friends know about the things she deals with so that nobody can judge her or her family. I see that she has given up on her siblings and doesn't even acknowledge her father who has abused alcohol her entire life and continues to do so today. I think we should never give up on our loved ones. We can always give it just one more try.

Randy Wood

There are legal consequences if you are caught using and abusing drugs and alcohol.

People may not think of the consequences of using drugs and alcohol because nowadays people glorify it. It's being praised everywhere from T.V., magazines, movies, and rap videos. In the movies, you often only hear of the fun that people have from controlled substances - like when it's a really boring party and no one is having any fun, it's only when someone brings some weed to the party that everyone starts to have fun and it turns out to be the greatest party anyone has every been to (until someone gets into a car accident or something).

However, there are legal consequences that can change a person's life forever. Being caught by police officers in possession of drugs and/or alcohol will get you in trouble. If you do not think so, then you are fooling yourself. Here are several examples of the consequences of "petty" drug use.

According to the Effective Solutions for the Texas Criminal Justice System, current law authorizes incarceration in the county jail for up to six months for any person who possesses two ounces or less of marijuana! Since Texas has the highest arrest rate for "petty" crime, I did some research and found that "57% of all 2003 drug possession arrests in Texas were for marijuana- almost half of which were young people between the ages of 15 and 25." A lot of people go to jail for drugs because police do not take drug charges lightly.

A former college sophomore named Isa just barely got his sentence reduced to parole after his father begged Judge Truman Morrison III to give his son parole instead of 45 days in jail. Isa still did not get off easy though; he had to complete 300 community service hours while on parole, and he was expelled from his university after his father worked so hard to get him in. In addition, if he violates his parole, Isa will have to serve 180 days in jail. All this happened because someone found out that Isa was selling drugs at his university, and when police officers went to investigate, they found a gallon sized zip lock bag full of marijuana, $3,000 cash and two digital scales. Now I am sure Isa thanks his father every day, and I think that Isa will never touch any kind of drug ever again.

I wanted to find out about how much trouble someone could get into for bringing drugs and/or alcohol to school. I looked at the Federal Drug-Free Schools and Communities Amendment Act and found that a single incident of possession of marijuana for personal use could get you a fine up to $100 and suspension. And for the people who bring alcohol to school, you could get a $50 fine. In addition, if you are underage your parents will be notified.

Now I want to talk about one of the most disgusting things I have ever heard of. Recently in Fort Worth, Texas, police in search of stolen goods got a search warrant and raided a house. The police found a video camera, which showed 17-year-old Demetrius McCoy and his 18-year-old friend trying to get McCoy's five-year-old and two-year-old nephews to smoke pot. The video shows the two little boys smoking pot like they've done it a hundred times. What makes things worse is that the teenagers were laughing as the two little boys stumbled and fell around and McCoy is their uncle. That right there is just horrible! I mean I am an uncle myself and I do not want to see one of my nephews or my niece ever even looking at any kind of drug! Now just because these two little boys have an irresponsible uncle, they were taken away from their home. Third degree felony charges have been filed against the teenagers. I think that they got off easy.

There you have it, people. Some of the trouble people could get into for using drugs and alcohol. After I have told you about all of these crazy and informative stories, I really hope you, the people reading this chapter, think about the consequences using of drugs and alcohol.

Sources

Olsen, Harald. "Sophomore avoids jail time for marijuana dealing."

The GW Hatchet Online. 1 January 2007. The GW Hatchet. 11 March 2007. http://www.gwhatchet.com/home/index.cfm?event+displayArticlePrinterFriendly&uStory_id

"Support HB254. Make small-time marijuana users pay, and learn." Effective Solutions for the Texas Criminal Justice System. 9 March 2006. Effective Solutions for the Texas Criminal Justice System. 11 March 2007. http://www.solutionsfortexas.net/id297.html

"Video appears to show brothers, 2 and 5, smoking pot." CNN.com. 5 March 2007. CNN. 11 March 2007. http://www.cnn.com/2007/US/03/04/pot.kids/index.html

"Federal-Free Schools and Communities Amendment Act." Eastern. Eastern Illinois University. 11 March 2007. http://www.eiu.edu/directives/drug/php.

Discussion Questions:

1. Why do people use drugs and alcohol even though they know the consequences of using those illegal substances?
2. What would you do if you knew someone who was addicted to drugs and/or alcohol? Would you say something or would you keep it a secret if they asked you to?
3. How do you think people who abuse drugs and/or alcohol feel about themselves or their situations?
4. Are you more likely to use drugs and alcohol if people in your family use them?
5. What steps would you take to help someone who was abusing drugs or alcohol?
6. If you were tempted to use drugs and alcohol, what would you do?

Why do teens start using drugs and alcohol?

Teens aren't born drug addicts; they become drug addicts. Many drug addicts start using drugs with their family and friends at a young age. It's acceptable, at first, when we are enjoying ourselves on holidays and special occasions. But, the next thing you know, you're doing it on the weekends and when you're bored. Before you know it, you're an alcoholic or a drug addict. In this chapter, you will find different perspectives on how people get hooked on drugs and the problems they face trying to stop. We interviewed people who have personally gone through these problems, people who have informed opinions.

Britta Ruona

It all begins at home.

According to the Greater Dallas Council on Alcohol & Drug Abuse, 65% of teens get alcohol from family and friends. If family and friends are the people who care the most, why are teens consistently getting such a harmful substance from them? Is it because their friends and family don't care, or is it because most people think that alcohol is not as bad for their health as most drugs? Well the fact is that alcohol kills 6.5 times more youth than all other drugs combined. Don't get me wrong, because drug use is also very dangerous, but more teens consume alcohol than drugs. They do not think that it will harm them as much as drugs will.

Studies show that by the eighth grade 52% of adolescents have consumed alcohol, 41% have smoked cigarettes, and 20% have used marijuana. The average age that an adolescent starts drinking is 11 for boys and 13 for girls. Studies also show that when parents talk to teens about the dangers of drugs, they are 42% less likely to use drugs.

I was surprised by some of the statistics. I would have thought that more teens get alcohol from friends and family than 65%. From what I have seen in my community almost all teens that drink or have ever gotten drunk get alcohol from friends or family. Also, I did not expect that when parents talk to teens about drugs and alcohol that they were 42% less likely to use them. I thought that those talks and conversations did not affect teens like myself because I know that when my parents talk to me about issues like this it goes in one ear and out the other. Most of the time I feel like I have something better to do and I already know what they are telling me. I never thought that it would make that much of a difference.

All of my research made me curious about the topic. I know a lot of teenagers who use drugs or alcohol but I never thought of the million-dollar question, *Why do adolescents start using drugs or alcohol*? I decided that I wanted to talk to someone who had a lot of experience with teenagers who abuse drugs or alcohol. I interviewed Ms. Lupow, a guidance counselor from my school, Bayard Rustin Educational Complex.

"I feel sad for kids who get involved this young because it's bad for them. I want to figure out why they're doing it," she said to me when we met. When students come to Ms. Lupow, she tries to do a number of things to help them. She will explain how bad it is for their body. She will also refer them to an agency that will help them. If the student feels comfortable, she will call their parents to get them help. After many experiences with teens involving drugs and alcohol, she has found that most of the time teens start because of one of two reasons: problems at home or school, or peer pressure. She also said that some teens look up to what they see in the media. For example, a lot of rap stars talk about drinking or drugs in their music. She thinks that schools should take more responsibility by going into the classrooms and talking about it. She said that presenters with past experience should come and speak to students because it would change students' attitudes towards substance abuse.

Before conducting this interview, I really did not think about the causes of teens using drugs and alcohol, only the effects. After this interview, I thought that if there wasn't a cause then there wouldn't be an effect. Even though I have seen teens get peer pressured into doing drugs and alcohol, I never thought that that was one of the main reasons teens start using. People with authority should follow Ms. Lupow's example and focus on how to prevent teens from starting drug and alcohol abuse. As Ms. Lupow says, "I want teens to know that there is help out there."

SUGGESTED RESOURCES

Books:

True to the Game by Teri Woods

Go Ask Alice by anonymous

Smack by Melvin Burgess

Teen Drug Abuse (Opposing Viewpoints) by Pamela Willwerth Aue

Forged by Fire by Sharon M. Draper

Movies:

Thirteen (R)

Holiday Heart (R)

Requiem For A Dream (R)

Girl, Interrupted (R)

TV Shows:

The Boy Who Drank Too Much

Websites:

www.drugaddiction.com

Christopher Then
To fit in with friends and family.

According to the American Academy of Child & Adolescent Psychiatry, teenagers may be involved with alcohol and illegal drugs in various ways. It mostly happens during adolescence because adolescence is a time for trying new things. There are many reasons why teens do drugs or alcohol. Some examples are curiosity, because it feels good, to take away stress, and to fit in with others around us.

This got me thinking that there have to be other reasons too, such as family problems, depression, and low self-esteem. Unfortunately, addiction is not that easy to give up. People often have to go to therapy. That's a hard commitment to keep. There have been lots of cases where people stop going to therapy and go back to their old ways. My cousin had a hard time dealing with this. He always used to drink alcohol, but then he realized that it was creating problems in his life. He tried to stop, but he couldn't. He says that it was hard to give it up once you are hooked on it. He also said that teens shouldn't start drinking or do drugs at an early age because they will keep on the addiction.

Often, when teens are drunk or high, they do not have as much control over their actions. And sometimes when they are really drunk or really high, they will not remember what happened. The worst thing is that their health deteriorates. There have been lots of scenarios where teens have gotten really sick because of these mistakes they have made. They hurt their bodies, but it seems to me like they don't care. So why do teens start using drugs and alcohol if they have seen so many bad results?

"A lot of them do it for many reasons, and the rate has increased recently," said Annette. She is 21 years old, and she has lost many family members and friends to drugs and alcohol. She feels that teens drink or do drugs at a young age because they think it's cool and because they are trying to fit in with others. She admitted that she has gotten drunk with friends and family members at parties, but she's said that she's not addicted to it.

I asked her if there are any alcoholics in her family. She said yes. She said they started at a young age and that they drank out of pressure. "If you didn't drink, you weren't a man," said Annette. She went on to say that her uncles put pressure on the young people in her family and she knew that was wrong. She turned them down politely, but sometimes she consented to please her family and friends.

That's the real reason behind the addiction. Some teens have tried to stop, but every time it comes back up. I think Annette and I have the same perspectives on why teens do drugs and alcohol at a young age.

After I interviewed Annette, I wanted to research more about addiction. I understood what she was trying to say, because I know tons of teens that have made the mistake of becoming addicted to drugs or alcohol. They know it's bad, but it's really hard to give up something that you are so used to, no matter how hard you try to stop. But it would be better if you heard their part of the story because I'm only saying one side. And to every story there are many sides.

Makera Watson

Sometimes medically prescribed drugs lead to worse things.

Alcohol and drugs are your problem, too. Alcohol and drugs are everyone's problem. There are many negative effects from the use of drugs in the society we live in today. Numerous people view drug abuse and addiction as society's biggest social problem.

According to the web site www.drugaddiction.com, statistics show that in most high schools in the United States, 43.3% of students drink alcohol and 20.2% use marijuana. The percentage of alcohol use was higher in Hispanic students than in any other race. 46.8% of Hispanic males and females use alcohol. 46.4% of whites used alcohol, and 31.2% of blacks used alcohol. 3.4% of students had used some form of cocaine (powder, crack or freebase) and 2.1% of students had used a needle to inject an illegal drug into their bodies one or more times during their lives.

Many people have questions about how and why these addictions begin. I have one answer to these questions because I have observed my drug-addicted grandmother. My grandmother started using drugs when she was hospitalized in her early twenties. She became dependent on painkillers. But it didn't stop there. In her thirties, she started using crack and cocaine, too.

While using these drugs she became pregnant three times, putting her kids at risk. (Fortunately, her kids turned out okay.) It wasn't until one of her children got a hold of the painkillers and ate them and was in a coma for several days that she finally stopped using crack and cocaine. But to this day she still uses painkillers, which made her lose lots of weight. She has lost her teeth as well. She gets cranky and vomits. She keeps saying she has pains so her children can send her to the hospital to receive more painkillers. Being with her is no fun. She nods off when she's outside or when she's eating, drinking, or doing her hair and nails. She sleeps all day and can't hold a conversation. She can't walk and she's hunched because the painkillers have weakened her bones.

Seeing family members and close friends use drugs has drastically affected me. I've seen how drugs can make people go through mood swings, lose good jobs, and suffer the risk of overdoses and suicide. The drug addiction website explains more of the negative consequences. For example, many people who use heroin feel paranoid and pick at their skin, making it bloody and sore. Some people that use PCP or "angel dust" believe they can fly, and end up jumping off of buildings and out of windows. Cocaine can make users aggressive and paranoid, causing them to become abusive in many ways. Ecstasy can cause you get an increased sex drive, and when using Ecstasy constantly, you begin to damage your brain and skull.

Watching my grandmother makes me realize that drug abuse can be really awful. It makes me feel like this is not something that I want to be a part of my life.

Source

"Why do teens get addicted?" Drug Addictions. 8 February 2007. <http://www.drugaddiction.com

Discussion Questions:

1. Were you shocked by the statistics you read about teen drug and alcohol abuse? Why or why not?
2. What advice would you give to teens about using drugs and alcohol?
3. Do you know anyone who uses drugs and alcohol? Is the person addicted? How might you help this person with his/her addiction?
4. What is your personal opinion on teenage drug and alcohol use?
5. How do think most addictions start? Why do teens feel the need to use drugs and alcohol?

Chapter 6

Depression

How do you help someone with depression?

In this chapter, you will learn about depression and how it might affect the people around you and, also, how people struggling with depression might be helped. If you feel depressed, you will find out ways to help yourself. If you have never been depressed in your life, you will get the chance to understand depression and help friends and loved ones who have it. You will also hear from other adolescents about how depression affects them and how being depressed has changed their lives.

Samuel Cuenca

By listening to them, supporting them, and getting informed.

"Depression is like reawakening someone's dark past or nightmare." These are the words of Mario, a college freshman who told me his thoughts on the difficult subject. Mario explained to me that, in high school, he saw many young adults and teenagers depressed about certain problems such as losing loved ones and living in poor conditions. Mario has lived through what my family goes through everyday because he is my cousin, so it is natural to relate to what he says. I asked Mario what he thought was the best way to deal with depression. He explained that, "The best way to help someone with this is by both comforting them and giving them their space." There are many other ways as well. You can try to talk to and listen to the depressed person. Encourage them to get involved in positive activities and to take care of themselves. Try to be fair when dealing with a depressed teen. If these small steps don't help, check some websites like webMD.com or

the U.S. Department of Health and Human Services.

Mario explained to me, "Some depressions are too great and they end up controlling the person." He believes this. He told me a story about one of his friends who was so deeply depressed after the passing of her mother that she killed herself. Personally, I believe the best way to help someone with depression is by getting to the source or the root of the problem. Some depressions do go as far as suicide. I believe that, if we are the friends of people with depression, we should help them prevent it from getting this far. It's like what Mario said, "I knew a lot of people who were depressed and they did many strange things to themselves, but you can't blame them. It's not them."

With all this depression, how does one deal with it? I recently found some information at TeenDepression.org that really caught my eye. Did you know that about 20 percent of teens feel depressed before adulthood? Between 10 and 15 percent of teens have some symptoms of depression at any one time. 8.3 percent of teens suffer from depression for at least a year at a time. Shocking, isn't it? It's important to help a fellow friend rather than watch him rot away in his own depression.

Source

"Teen Depression Statistics." TeenDepression.org. 2005. Teen Depression. 2 February 2007. http://www.teendepression.org/articles5.html.

SUGGESTED RESOURCES

Books:

Girl, Interrupted by Susanna Kaysen

Chicken Soup for the Teenage Soul on Tough Stuff by Various Authors

Undoing Depression by Richard O' Connor

Teen Depression by Michael. J. Martin

Now You See Her by Jacquelyn Mitchard

I Never Promised You a Rose Garden by Joanne Greenberg

Prozac Nation by Elizabeth Wurtzel

Movies:

Real Life Teens: Teen Depression

Everyone's Depressed

Girl, Interrupted (R)

Melbin Peralta

By finding professional help.

"I have been depressed and it feels terrible. There's too much to handle and every little bad thing that happens just adds to the situation and you feel as if you hit rock bottom," my friend, Michelle, said as she responded to one of my questions. The more we experience in life, the more we learn how strange life can be. There are ups and downs, from falling in love to losing a loved one, from having a job to losing it in a second. Those are the reasons some people feel depressed. Michelle's quotes are true because I've experienced how it feels to be depressed. I truly believe that man's worst enemy is himself. It is true because sometimes our decisions can have an effect on us. For ex-

SUGGESTED RESOURCES

Music:

"Oh Mother," "Hurt," and "Beautiful" by Christina Aguilera

"No More Drama" by Mary J. Blige

"Down" by Rakim Y Ken-Y

"Boulevard of Broken Dreams" by Green Day

"Runaway Love" by Ludacris featuring Mary J. Blige

"Lonely" by Akon

"Can't Take That Away" by Mariah Carey

Websites:

TeenHealth.com

webMD.com

mayoclinic.com

ample, we hurt our loved ones by committing errors which could lead to depression. This is a very touchy subject, especially when family or friends are the ones affected.

Sometimes we all, as humans, experience moments when we're depressed. When this mental disorder strikes, however, often there is a huge problem. "I think depression is a disease. I think it is an illness because it is either that something tragic happened in your life and you're really, really sad or it's that you're clinically depressed," said Michelle. "When you're clinically depressed, sometimes you have to take pills and it's only mental." According to Eli Lilly and Company, depression is a common mental disorder that affects young teens and many adults. Some causes of depressions are sadness, loneliness, the death of a family member or even having family history of depression. Depression might affect someone emotionally, but it can also affect someone physically. Some symptoms include lack of appetite and even lack of energy. "Depression affected my health by decreasing my weight," said Michelle about the symptoms she had. "I didn't have an appetite. I actually lost 20 pounds and I was always tired."

According to Eli Lilly and Company, thousands of American adults suffer from depression because of marriage problems, financial problems or moving away from love ones. But, on the positive side, depression can be treated with medication. Most importantly, your loved ones are there to help you. Also, doctors, psychologists, teachers, friends and family members can help. When you're depressed, realize that your friends or someone you can trust are there for you. "I didn't have anyone to help me because I didn't want anyone to help me," Michelle said about how she made a change by getting professional help. "But then I got help by talking to my counselor and friends." Always remember that you're not alone and, if you feel depressed, seek help.

Source

"Symptoms & Causes of Depression." Eli Lilly and Company. 2007. March.6, 2007. http://www.cymbalta.com/depression/understand/causes.jsp?reqNavId=1.3&ccd=mdddtc88&WT.srch=1

Fannie Ventura

By reducing the pressure to succeed.

Who said that teens don't suffer from depression? Well, many do and on a regular basis. Some teens are suffering from depression without knowing the reasons why. I thought that I experienced depression one time when I was dealing with the parents-meet-the-boyfriend thing and everything went so wrong. My mom told me to bring my boyfriend home for her to meet, but he didn't want to take that step yet. I became extremely sad because I had already told my parents that he was going to come over. Sometimes, we teens think we might be experiencing depression, but it's only a bad moment. My sad moment was temporary, but depression is different because it seems to have no end.

I interviewed my 19-year-old sister, Arlene, and she told me that she has struggled a lot with real depression. She said that at an early age, when she was about 10 years old, she was bullied at elementary school by other kids. She said that one time a girl just wouldn't stop bothering her at lunchtime. The girl started to pick on my sister because she was jealous of her looks. After that, my sister had to fight her because she was getting frustrated that people in school were bullying her for no reason. The whole situation made her feel depressed.

Another problem Arlene had that led to depression was stress from so much pressure at school. Now that she has entered college, she has to pay for everything like her tuition expenses, her monthly laptop bill, her credit cards and cell phone bills. She also gets stressed because she doesn't have enough time for herself and she gets mad that she has to pay for so many things at the same time. It's not easy being in college and having debts for other things and, on top of that, hearing your mom and dad say, "You need a job! Stop being lazy!" She feels down and depressed because of the pressure they put on her. It's a tough situation for her. At times, she thinks to herself, what it would be like if I didn't have all these problems? She knows that she gets depressed by the way she feels emotionally and by how she feels about what people think of her.

On top of all this, Arlene has asthma. Sometimes she starts wheezing, and then she cries. She feels down, without even the energy to get out of bed. She does exercise but her body feels weak at times. Depression for her is like a road without end; it just keeps on going and going. She thinks that depression is too much for her to deal with.

Teen depression happens to many people and we know it is hard. Sometimes teens decide to cut themselves, try to get in trouble, or try to get rid of their problems by doing bad things. But I feel that depression is a real expression of how you feel and a result of pressure people put on you. There are many ways people use the word "depression." Some people think it is stress; others say depression is trying to get rid of problems or obstacles in their way. My sister's situation shows one way that you can help someone with depression because she is an example of what depression is like and how she feels about it.

Roberto Vicente

Advise depressed people to seek out the help of trained professionals.

Once there was a girl who cried every night. Selena was her name. Selena is a high school student who is only 16 years old and who has already gone through depression. When I considered who could tell me the most about this topic, she came to mind because she's a friend and she was very open about this difficult subject.

Selena was depressed for a long period of time. When she was 13 to 14, she began to have many problems at once. "Life goes by so quick and I just wasted a big part of it," she said. I asked her how she dealt with her depression. She told me that, at first, she locked herself up in her room and just came out to take a bath. She didn't even eat for days. She cried so much and all that came to her mind was darkness. She said she wondered what the point was to live such a hopeless, difficult life.

Soon her mom noticed that something was wrong—she was not eating, and she was not her usual happy self like she was as a child. It took her mom about a month to see that her problems were really serious. She took Selena to a psychologist right away, not knowing what was troubling her. It was so hard for Selena to admit to her psychologist that it was depression and not just bad feelings. The truth was that she didn't want help at all. She just wanted to be alone for a while.

Selena was so torn apart inside that she needed a psychologist in school. After a short period of time, she thought that it was going to make her feel better. Maybe it did, but everything came back after she stopped going. She did not cry as much but she couldn't feel any worse.

Finally, after five months, Selena realized, "This wasn't good for anybody and I was wasting my life. I started going out more, I went to school more often, I ate more so that I wouldn't be weak, and I actually started talking to people so I wouldn't feel lonely." Her friends helped her out with support and advice. They told her how bad it would be if she just quit. They encouraged her to be strong. They told her that everything was going to get better soon.

I believe that many people go through depression at early ages. A family member, who is eleven, is also going through depression. Sometimes I believe that I also have it. I've been down for stupid reasons or no reason at all. Knowing that being down is a symptom of depression, I have tried to snap out of it by going out and talking to my cousin about how good and bad life is. But I wanted to find new ways to help people with depression.

To find out, I researched www.thechangeyoudeserve.com for ideas. One suggestion I found is to encourage teens to get therapy. If you know depressed teens who refuse to get help, you should seek out a professional. If teens get therapy, it might help them to understand why they are depressed. I think that once you're depressed, you want to be with people who understand you.

Teens can also learn how to control or deal with their depression through medication. Medication prescribed by a psychiatrist might be necessary for the patient to feel relief. I personally believe that they are not good. A family member has depression and his medications never seem to work. Sometimes I think that they only make matters worse because once the person stops taking the pills, he or she might revert back into their depression.

Depression is usually caused by more than one problem or situation. That's why it's so hard to get rid of. Anyone with depression should try making new friends, perhaps by participating in sports, jobs, and school activities. Staying busy helps teens focus on positive activities rather than negative feelings or behaviors. When problems are too much to handle, you should try to ask a trusted friend or adult for help. "No one should be afraid to ask for help," I said to myself after reading all about depression.

Source

"Depression in Teens." NMHA. 2007. Mental Health America. <http://www.nmha.org/go/information/get-info/depression/depression-in-teens

Discussion Questions:

1. What do you think are the best ways to deal with depression?
2. If you were giving advice to someone who was struggling with depression, what would you say?
3. What are some of the reasons people experience depression?
4. What effects does depression have on a person's life? How might you tell if someone is depressed?
5. If a friend suffered from depression and tried to hurt herself or others, would you tell an adult even though your friend asked you not to?
6. If you were depressed, what would you do? Who would you go to?

Why do teens commit suicide?

Depression is a major issue these days. More teens are suffering from depression and are looking to suicide as an answer. Knowing this, we wondered why teens feel that there is no way out other than suicide. We interviewed people to try to find some answers.

Kevin Diaz

Teens have a lot to deal with but should try to focus their attention on positive things.

We all have points in our lives when we just feel really sad. Things happen and sometimes there's just nothing you can do. You would be surprised by the amount of people who have experienced deep sadness. Ms. Sinclaire, our student teacher, gave me an opportunity to find out more about teen depression and her perspective on it. She told me that many of her friends went through depression back in her teen years in the mid 1990s. I was shocked when she told me depression was "popular" at her school. Many teens in her neighborhood thought that it was cool to be all sad and depressed. They were dealing with family problems, school issues, and relationship problems such as bad break ups or arguments. In contrast, Ms. Sinclaire was a happy teen. She would rarely feel depressed or sad. Everyone else would just live their dark, cold lives. She didn't pretend to be depressed to try to fit in. This made her seem unique because she was the only teen in her neighborhood who would rarely feel depressed or have the urge to commit suicide. A lot of teens resented her because of that. The reason why she was always happy while other teens weren't was because she would try to move past her sad feelings. She never allowed any type of sad or suicidal feeling to enter her heart.

It's amazing how teens have always dealt

with inner turmoil and struggles. In general, I think it's always a struggle for teens because, at our ages, life really hits us hard. We are not really children anymore, but we don't have a sturdy heart like adults do. We often can't handle the pain in our hearts, and that leads to depression because we just don't know what to do with our lives. It takes us time to figure out life and sometimes it seems like we won't ever figure it out so we might just want to commit suicide. Suicide can seem tempting for many different reasons. For example, we teens have our relationships but sometimes terrible breakups occur and we just want to kill ourselves because we might feel like we've lost our entire world. I think suicide comes from teens who never took a chance to move on in their lives.

I have to give all my thanks to God for creating me with such a strong heart; sadness never even approaches it. I'm always a proud person, especially for my age. Sometimes, I might feel just a little bit too proud. There are a couple of reasons why I always stay joyful. One reason is my family. They are what I love most. They are always in great moods and they are really funny. Another thing that keeps me proud is baseball. I live for that. I always stay focused on it, no matter what, because I know it may have a big impact on me in the years to come. I always keep my hopes high because I know I'm young and I have a bright future.

Going back to Ms. Sinclaire... She sets a great example of a young teen trying her best to move past the depressing situations that go on in life. Whenever she would begin to feel depressed, she would go out and have fun or turn her attention to something positive. That's what teens should do, if possible. That's the best way to forget about depression. Just hang out, laugh, smile, and enjoy life any way you can. We teens have to look ahead and expect the best from life. Teenagers should move their attention somewhere else when feeling depressed. We need to try to see the big picture. As Ms. Sinclaire said, the older you get, the more you will understand that life goes on and there are many opportunities to change your life for the better.

SUGGESTED RESOURCES

Books:

Lisa, Bright and Dark by John Newfeld

Stay With Me by Garret Freymann-Weyr

Tears of a Tiger by Sharon Draper

House of the Scorpion by Nancy Farmer

Chicken Soup for the Teenage Soul (All Editions) by Jack Canfield, Mark Victor Hansen and Kimberly Kirberger

The Power to Prevent Suicide: A Guide for Teens Helping Teens by Richard E. Nelson PhD, Judith Gales and Pamela Espeland

Movies:

Romeo and Juliet directed by Baz Luhrmann

Websites:

www.gurl.com

Cynthia Rodriguez

Teens sometimes consider suicide when they feel overwhelmed and alone.

"Suicide is stupid… You should live until it's your time to go, not try to change your fate." This was said by a person who has dealt with suicide before. He has a lot of opinions and many experiences and, when it comes to suicide, he could help a person who needs it.

There are a lot of different reasons for suicide, but not many people to talk to about it. Alex, a junior in high school, believes that life situations might cause depression, such as feeling like you're an outcast—specifically if you're homosexual—and having relationship issues. It seems that there are not many people, not even adults, to help you with these issues. To Alex, teachers, guidance counselors and other adults in general are hard to talk to about suicide because they often make assumptions. They assume that we might hurt ourselves if we bring the subject up, then they put us in situations that we don't want to be in. A person dealing with depression might end up thinking, "I'm going through a lot of things that you can't help me with," as Alex says.

The more I spoke with Alex, the more he helped me to understand what he had been through. He has attempted suicide before. Things weren't going right. He couldn't keep a steady relationship and there was no one around. People always called him with their problems but they would never listen to his. He was always partying to hide his loneliness. In his depression, he drowned himself in marijuana and alcohol and would wake up not remembering things. Alex attempted suicide by slitting his wrists and jumping in front of cars. Eventually he "hit rock bottom." He broke up with his girlfriend and went to Florida where he quit drinking and smoking and had a talk with himself. He told himself that if he kept going at the rate he was, he was really going to end up dead before he turned 21.

How would you help someone who is suicidal? It's really tough to give helpful advice to a person who is thinking about committing suicide. Alex says that he would just tell them not to do anything stupid. "You just end up suffering," he says. Going through depression and having thoughts about killing yourself is serious and people who do need help, not only because they're harming themselves but because they're harming the people around them. One of Alex's friends wanted to tell him about his own problems, but it turned into an argument. Alex was going through his own depression, and he just didn't want to hear anything from anyone.

Alex's point of view is an interesting one to me, but how I feel is somewhat different. There are always reasons why people feel suicidal, but not everyone can understand why. I think that when a person is experiencing suicidal thoughts, choosing your words correctly is important. Any wrong word might either cause an argument, or even cause the person to actually do what you're trying to get them not to do. Dealing with depression on your own is not always a good thing so try to confide in someone you really trust. If it helps, it's not a waste of time. But I do agree with Alex that there are times when all you can do is just sit down and really think about what's going on in your life, like he did.

Karen Rosario

Because they don't always know about or seek out the help that is available to them.

Depression affects teens all over the United States, and all over the world, in many different ways, which makes it a very complex disorder. Some adolescents might have severe depression; others might endure mild depression. But all of it is detrimental to the person experiencing it.

I interviewed a school psychologist to get his perspective on this important issue. He has been a psychologist for 37 years and says, "Depression is a very dangerous disease that can kill people who suffer from it because they can commit suicide." This illness can first show itself in many different ways. For example, the depressed person may experience problems with daily activities, such as eating and sleeping. Even though depression is very common, the school psychologist reiterates that the causes are unknown. Some believe depression is caused by a chemical imbalance in the brain that affects the nerve cells so depression isn't just imagined. It's real, it's scientific, and it can happen to almost anybody.

The next person I decided to interview was a classmate, Chrystal. She has been depressed in the past and it lead to her cutting herself, taking pills, drinking, and even trying to commit suicide a few years ago. Chrystal once wrote a suicide note to her mom saying that there was only one thing that would cure her depression—and it wasn't what but whom. Those people were her little cousin and her best friend. Her younger cousin at the time was eight and had lost her mother two years earlier. Her best friend was always there for her. Chrystal began visiting her cousin almost daily to talk about her situation. It was like therapy. Once, after a discussion, her cousin said to Chrystal, "I wish you were my mother." This meant so much to Chrystal because there was now someone who needed her to live. Furthermore, her best friend opened her eyes to all that was worth living for.

All of us have times in our lives when we experience the symptoms of depression, like sleepiness or lack of appetite, but that doesn't necessarily mean that we're depressed. Those symptoms can be treated through eating well, getting exercise and, most importantly, feeling loved and supported by loved ones. If symptoms persist over a long period of time, people should seek medical or professional help so they don't have to deal with it alone.

There are many ways you can help a teenager deal with depression. There are numerous public hotlines open 24 hours a day, seven days a week. You can also take a friend to a clinic or a hospital. If that person is trying to commit suicide or hurt someone else, you should get help from an adult immediately. If you don't speak up, then somebody might wind up seriously injured or dead when all of it might have been prevented.

In my opinion, depression should be treated as promptly as possible, whether through therapy or medication prescribed by a doctor. If a depressed person doesn't get help, the consequences could be serious, and that really concerns me. I, myself, used to be depressed and it is not easy to deal with, but teens should know that they deserve to enjoy their lives. Depression is horrible but, with the right help and support, it can make you stronger and, like me, you might love yourself even more.

Discussion Questions:

1. What kinds of experiences might lead to depression?
2. What would you do if someone you cared about was going through depression or wanted to commit suicide?
3. Why do you (or don't you) think it is important to talk to someone when you're having issues or thoughts about suicide?
4. Is this problem more common than you assumed? If so, why do you think that is?
5. In your opinion, do you think the medications used to help depression are effective? If so, why? If not, what do you think would be a better way to help people with depression?

Chapter 7

Family Life

Are traditions important?

Many people believe that tradition and religion are the same while, in fact, this is not true. There is a major difference between the two. Traditions are similar to habits while religion is more about strong beliefs. We discovered interesting facts about tradition through research and interviews with different people. The following pieces represent points of view from different perspectives. We decided to show two very different points of view on tradition to broaden the reader's understanding, and we wanted to make the reader think about what tradition is and how much it affects his or her life.

Liliana Perez

Traditions are important because they bring families together and help you form your identity.

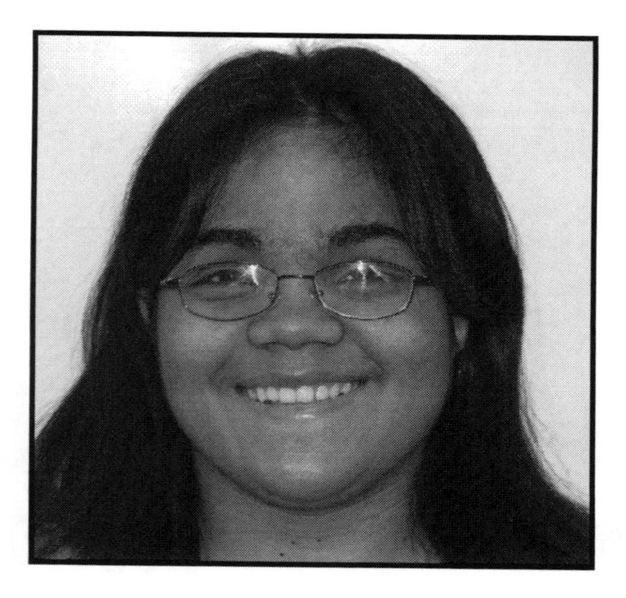

Traditions mean different things to different people. When I used to hear the word tradition, I thought it was just a fancy word for religion. Through researching and interviewing, however, I learned that tradition means more than just religion. During an interview with Kevin, a student at my school, I discovered that his way of understanding tradition was through family and religion. He voiced it as "generation after generation of tradition." Christmas, get-togethers, and other forms of family-oriented events are common in Kevin's family. These times are filled with dancing to Spanish music, eating food, listening to family stories, catching up on old times and learning about the new.

During these times, they feel safe together. These are constant traditions that they never stop doing because they don't want to lose the tight feeling of family. Kevin grew up among family, so he took the word "tradition" to mean that.

While reading "Family Traditions Are Important" by Marsha K. Weaver, I discovered a different way that tradition can be viewed. She wrote, "Traditions are the thread of life that bind us together as a family." She also states, "Families that have the strongest bonds have the most tradition or rituals in their lives." She believes, as I do, that these traditions or rituals create a sense of continuity, understanding, and love that strengthen the family's closeness. Tradition cannot be specifically defined. Rather, it is something that is experienced on a very personal level by different people. Traditions in a family help us appreciate each other and keep close to our relatives, both past and present.

Traditions also make a family's uniqueness shine more. For example, you may eat a certain food or celebrate a certain holiday that another family may not. When my family gets together, we do similar things to Kevin's family but we do not celebrate Christmas or any other holiday. To make up for the things we don't celebrate, my family has more get-togethers or we celebrate little things like baby-showers, graduations (we hold graduation parties for the little ones in our family, too, so they don't feel left out), pool parties, and anything else we can think of (even if we have to make it up). I'm sure your family is different in their own way, and that they celebrate what is important to them. This makes your family unique, which should make you appreciative of them.

Traditions can also be a learning tool. When you were younger, I'm guessing that you asked questions about everything. The older generation of your family most likely told stories about why they celebrate the things they do and why they eat certain foods. You may have grown up on family stories and get-togethers or some form of family togetherness. Perhaps you learned to love your family's craziness, good and bad, through tradition. You learned who you are and about your family's history. You learned to love your cultural identity through tradition. The traditions in your family made you a stronger person, mentally and emotionally. These are tools that are very important in every person's life, ones you can't learn in a textbook. You may learn the word "tradition" in a textbook but not the importance of it. Tradition, as you just learned, is more than religion. It is a very important tool that can help to make a family stronger today and, also, to create a stronger bond for the future.

SUGGESTED RESOURCES

Books:

The Chosen by Chaim Potok

China Boy by Gus Lee

The Woman Warrior: Memoirs of a Girlhood Among Ghosts by Maxine Hong Kingston

How the Garcia Girls Lost Their Accents by Julia Alvarez

A Tree Grows in Brooklyn by Betty Smith

Movies:

Bend It Like Beckham

Soul Food (R)

My Big Fat Greek Wedding

Fiddler On The Roof

The Prince And Me

TV shows:

7th Heaven

Ugly Betty

The George Lopez Show

Jenna Tin

Religion is a tradition that gives you the guidance and support of a community.

Religion is a pretty extensive subject, and it's one that many people might not feel comfortable talking about. I, personally, am not religious but my family is very traditional and, in a way, I do feel that there's a connection between the two. From my perspective, tradition seems like a hassle at times because I'd rather hang out with friends or I simply have something better to do. Because I'm not religious, the variety of religions and the differences between them all seems very complicated. It's relatively baffling to me how so many personal conflicts and worldwide disagreements can result from something that should ideally unite people instead of separating them. A conversation with my English teacher, a woman who considers herself "culturally Catholic," opened my eyes on the true meaning of religion and what it's like to be religious.

Ms. Quigley grew up in a religiously traditional family. She attended Sunday Mass, Catholic school, studied the Bible, prayed before meals, and celebrated all religious holidays. The community she lived in was also considerably religious. Although she did not discriminate against those who weren't devout, she did think it was strange if they weren't. Looking back now at her childhood and teenage years, she describes vividly how church gave her guidance, comfort and support when she needed it the most. "When I was younger, I thought religion was a unifying idea that encouraged peace," she says. When she attended college away from home, she began questioning a lot of things—religion being one of them. She was very close with her family and felt extremely lonesome and secluded from them because they were all devoted Catholics. After college, she volunteered for the Peace Corps for two years. She was in Africa during the time of 9/11. She heard about it over the radio and was in disbelief that this was happening yet she was entirely shielded from it. She says, "[Religion] can be used as a tool to divide people instead of unite them."

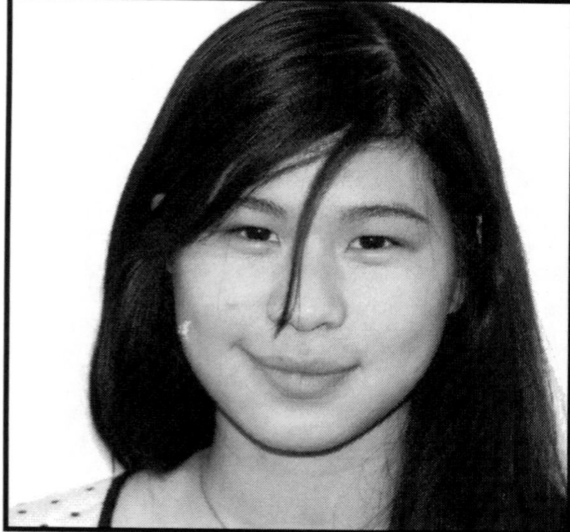

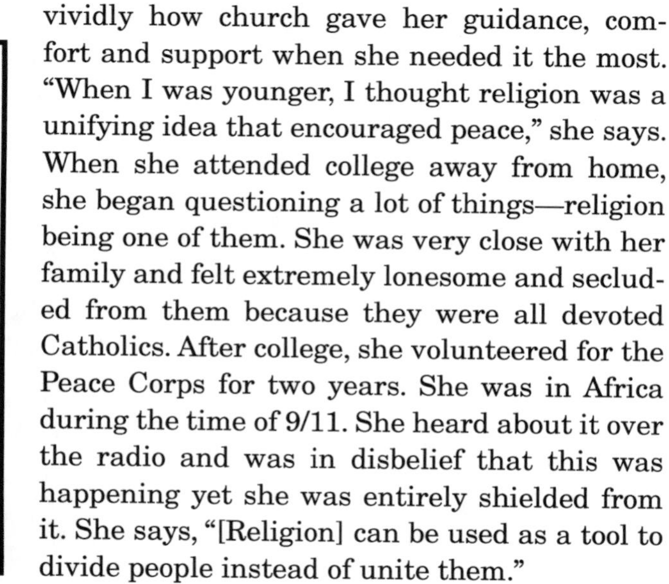

Ms. Quigley explains that one of her many experiences with religion was when she decided to get married. She was afraid that her husband, Chris, a non-practicing Lutheran, would not be comfortable getting married in a Catholic church. She faced a huge personal and family debate. Many of her family members were so upset by this idea that some of them even refused to attend the wedding if it wasn't in a Catholic church. She and Chris attended marriage counseling with her priest and, soon thereafter, they got married. It turns out that the priest was pretty lenient with the wedding ceremony. They did get married in a Catholic church, but without some of the main Catholic traditions.

When asked what she misses most about religion, Ms. Quigley replies—with a distant look on her face—that church gave her guidance and a sense of self-improvement. "I felt comfort in religion. I felt like church was a community." Growing up now as a teen, I know life isn't easy—nor will it ever be. The world is one huge bumbling mess and I'm caught up in the middle of it. I can definitely see how religion played such an important role in Ms. Quigley's life. It's nice to know that whenever you feel out of place, there's always somewhere you can feel at ease. My only objection is that religion should be practiced freely and people shouldn't be judged for their perspectives. As Ms. Quigley states, "Anybody should be able to believe anything they want and practice it any way they want, as long as it does not hurt others or put others down."

Discussion Questions:

1. What sorts of traditions does your family have?

2. How are traditions in a family helpful?

3. How are traditions in a family hurtful?

4. What are the effects of breaking your family's traditions? On you? On your family?

5. When you start your own family, what sorts of traditions do you plan to preserve or begin? Why?

Why do teens take on adult responsibilities?

In general, teens have overwhelming lives. With all the changes they face, they can feel pressured. Some of the things teens deal with include puberty, peer pressure, school, work and social lives. On top of that, some teens have to take on adult responsibilities, forcing them to grow up faster. They don't get to enjoy their adolescence like they should. In this chapter, you will read different perspectives on the hardships that many teens face today.

Yenedy Arias

Teens take on more responsibilities when parents act like children.

When teens complain that they are stressed, it's hard not to wonder what kind of lives they are living. I can't really compare my life with theirs when my friends talk about what hard-knock lives they have. I can't say I'm not happy in my home. My parents are always kissing and holding hands, and my brother and I have our own rooms. We each have a large bed, a TV, and a computer. We aren't living "the life" but we can't complain. We have basic chores: clean our rooms and the bathrooms, throw out the trash and do the dishes. A typical day in many of my peers' homes looks like this... they go to school and then home, take care of siblings and witness arguments. They have problems with their boyfriends or girlfriends, don't have enough space and often cry themselves to sleep because life is just one constant pressure. Since I'm the youngest in my family, I have fewer responsibilities, but I would like to know why teens are complain-

ing like they have a nine to five job and are getting minimum wage.

To me, going home is a time to relax, hang out and just be with the family. Every time I get home I get a good feeling, but that isn't the case for a 16-year-old named Andres who I met in Spanish class. He is a calm, peaceful and generous person and his tranquility drew me to him. Hanging out later in a quiet hallway in my building, he told me about sharing a crowded, two room apartment with the five members his family: his mother, father, grandfather, and younger sister. You might think that because he lives with both parents that would be something good but, in fact, that is precisely the problem. Andres' parents love each other but, for some reason, they end up arguing. He said, "They can start out with a usual conversation about what they did during the day and end up arguing about something that doesn't relate at all." Andres is the one to break up the fight. He screams at them, telling them to stop as if they were children. "I don't think I should be telling them to stop or shut up," he said with a frustrated look. They usually argue about how money is spent, rent, his sister's school problems, and needing space in their home.

Not too long ago, his father had a heart attack and now stays home all day. Recently, things have been worse because his father is depressed and is even more critical of his family. He collects random things that make the house seem smaller and smaller. He is short tempered, strict, and sarcastic. When he says something, that's what goes and when he tries to be funny, you'd better laugh. Andres used to stay home with his father and play video games. Now, he is in his first serious relationship and tends to go out more often. Andres said that recently his father said, "F--- you" to him, something he had never done before. Andres' voice sounded as if he didn't want to admit it. The family thinks it might be because his father is jealous that his only son is out and not with him.

His mother, on the other hand, is barely home. She likes to leave and visit her best friend's house so that she can get away from the fuss, the arguments, and bickering. Because his mom isn't around a lot, his sister looks up to him more. He gets annoyed but it's a younger sibling's job to bug you. He likes it when his mom is home because he feels like his mom puts more of an effort into caring about school. He says he feels like he is missing out on a lot when his mom isn't around.

He describes his grandfather as "the peace maker." He goes to him to talk about what goes on because he is a very good listener. You wouldn't want to argue in front of him because he is older and should be respected. "When I need somebody, I go to him," Andres told me. "He won't get in the way. He'll just give me advice." His grandfather is his guidance and helps him get through his days a little easier. Andres is usually stuck in the middle but claims that he can never rid himself of the arguments. His grandfather gives him what every neglected teen needs—that special someone to help him get through the hard situations in life—especially because teens can have such heavy burdens.

SUGGESTED RESOURCES

Books:

Maniac Magee by Jerry Spinelli

Oliver Twist by Charles Dickens

Lord of the Flies by William Golding

Little Soldier by Bernard Ashley

Children of Neglect: When No One Cares by Rowena Fong

Movies:

Newsies

Home Alone

The Outsiders

The Breakfast Club (R)

Platoon (R)

Raising Victor Vargas (R)

Real Life Teens: Broken Homes

Delia Calix

Teens take on more responsibilities because parents are working too hard.

"One effect of [teens taking on adult responsibilities] is that it takes away part of their childhood," said Dr. Monica A. White, a consultant at the Bayard Rustin Educational Complex and lecturer at Teachers College, Columbia University. "They lose their childhood innocence," she said in response to my question about what happens to those teenagers who have no choice but to take on the "parent role". "I think there's always too much. So much responsibility makes them sleepy in class. It affects social relationships and education. Children shouldn't take on adult roles."

Year after year, we hear more cases of teens taking on their parents' responsibilities. They are cleaning the house, making food, working, and taking care of younger siblings. Taking care of the household so much can consume their lives. Sometimes, parents are nowhere to be found and unaware of their actions. Some parents are having fun, working too much, or acting childish. Teens are doing so much to help their families that they can't enjoy themselves and have normal teen-age lives.

Sometimes, parents don't realize all the pressures they put on their kids. They are so focused on working as hard as they can to pay bills and buy food that they don't have time for their kids. There was a time when my parents were like that, not so long ago. Every single Monday through Friday at 3pm, I had to rush straight home from school to pick up my sister and baby brother. I stayed home to take care of them—to feed, teach and put them to sleep—and I was the one giving them rules and disciplining them. I love them but I felt trapped. Since I'm the oldest child, I had to help my parents. They would come home very late from work and I felt helpless to tell them anything, seeing how tired they were. As time passed, however, I was missing out on so many things that I wanted to enjoy and I started to feel rebellious. I wanted to get away and escape. I needed my space, my freedom. I tried telling my parents but they didn't under-

stand. I tried being mad at them and they would either get mad at me or act like they didn't care. So I reasoned with them. They would leave early for work around 7am and wouldn't come back until 11pm. I told them that it was affecting me and it was serious because it was depressing me.

It took a bit of time, but it finally worked. My parents are very thankful for all the things I've done and I'm grateful. Now, my parents have hired a nanny for my brother and they come home earlier so I get to relax and go out more. I'm doing better in school and don't feel the need to rebel against them, knowing that they don't deserve that. I love my parents very much and being honest has helped me so much. Finally, I can breathe!

Teens need to live their lives and have their own experiences. Adults should help them choose what's best for them. If a parent can't handle the job of parenting, then they might share it with another adult like a grandmother, aunt, or cousin. Parents should make sure that whatever chores or jobs they give their teenage children don't get in the way of their other responsibilities—like homework and school—or their social lives. Parents should encourage their kids to join after school programs or other activities. Of course, parents have rough times and they still make mistakes. Teens can help their parents, but they can't do everything at once. Remember, like Dr. White said when she quoted an old African saying, "It takes a village to raise a child."

Robert Montero

Parents put too much pressure on their kids.

Janet was having problems at home and I couldn't believe that she really thought about killing herself so that she could get the attention she deserved. It seems like everywhere you go—home, school, the streets—there are huge problems with kids raising themselves. Consequently, friends can sometimes get you involved in things that can get you into real deep trouble. If parents were around more often, this might not happen. Much of the time, the biggest problems begin when young people are being brought up without enough support and often without an adult figure to look up to. Sometimes, kids grow up with no parents at all and they have this feeling of being lost because they ask themselves, why don't I have a father or mother? Because of this, some kids become suicidal and feel like they have nothing to live for.

Janet has been a friend of mine since fifth grade and we've been really cool with each other ever since. She always tells me things about her personal life and it's often about her taking on the parent role. Janet is home alone a lot because her divorced parents are always working just to make ends meet. On top of this, she has to watch her little brother who's only a baby. She lives with her mother who comes home tired and doesn't have time to give Janet the support that she needs and deserves. Janet works very hard at school so that she can make a better life for herself. Unfortunately, her parents pressure her tremendously to do well and sometimes the cracks show. For example, her father, who never reached his own dream of becoming a basketball star, pressured Janet to excel in the sport. Janet ended up quitting because she couldn't take it any more. So much of her life is about pressure and so little of it is about being a happy, carefree teen.

For teens, life is really hard when parents don't get involved in ways that celebrate their children. Teens end up feeling like they have to create their successes on their own. This can lead them to try and find an escape from all the negativity. For some, that escape may be books or school, but for many others it's sex or drugs or gangs. They use drugs like weed to relax and get through the day faster. Some teens actually get into drug dealing to get money because their parents can't buy them what they want. Other teens, who have no support at home, may turn to gangs that will treat them like family and they can end up doing crazy dumb things. Those who can't find any support at all might resort to hurting themselves.

There's a lot to learn from Janet. Her mother gave her so many responsibilities that now Janet is more like an adult than a teen. I'm not a person who likes to give advice but, if you're a teen like Janet, try confronting your parents about it. And, if that doesn't work, make sure that you have as much of your own fun as possible in your spare time.

Danaya Williams

Taking on adult responsibilities prepares teens to become better adults themselves.

Ms. Outlaw is a social worker at Harbor Family Horizons located on 113th Street and Fifth Avenue. I interviewed Ms. Outlaw because she has had firsthand experience with what it is like to be a child who takes on adult responsibilities. She could relate to the topic a lot better than a teen who grew up with both parents present and took on the "normal" roles of a teenager. "About one-third of my cases have children who take on parental roles," said Ms. Outlaw. But exactly what are the normal roles of a teenager?

It's normal to see a sister picking up siblings from school or taking them to the doctor. It's just that, in the society we live in now, we're seeing the older sibling take on more responsibility. Is this a good or bad idea? Does this make teens more mature, or does it take away from the teenage experiences every teen should have, like a trip to the mall or first kiss? Well, Ms. Outlaw grew up as the eldest of three young girls. "I was spoiled by my grandparents," she said. Her mother was a hard worker who left her to take care of her two younger sisters. "I took them to doctors' appointments, helped them with homework, and did other things for them."

I, too, can relate to this topic. I grew up living with my grandmother after my mother died when I was young. I had two older sisters and a younger brother. After my sister grew up and left the house, I was alone and wanted to leave. My grandmother was getting sick and she needed a break from me so that she could recover. That's when the option became available for me to stay with my aunt and uncle. I loved the idea of being together in a house with other kids and the thought of a normal family, so I went through the process of adoption and I was theirs.

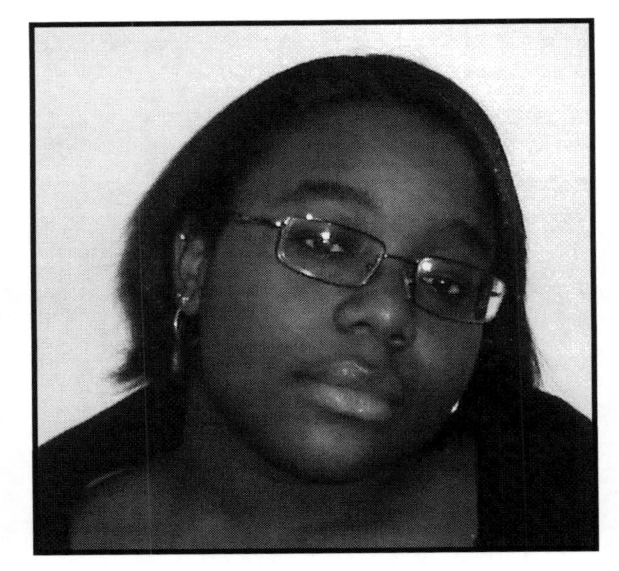

Life at home was so fun with me that they wanted to adopt more kids. I was ecstatic because I had always wanted a little brother to live with us. I jumped for joy and even gave up my room so that he could have a nursery. When he arrived, he was four months old. He was handsome and playful. As time went on, we grew accustomed to him. Our parents became comfortable leaving him with us, and that's when the responsibility kicked in. We wiped tears and changed diapers and gave baths.

Later, they decided to adopt his little sister so that they didn't have to be apart. She was only two days old when she came home to us and that's when my responsibility came into play. My parents are pastors who have long hours so they counted on us to step up. When I came home each day from school, I had to do my chores and my homework before my little sister came home so that she could have my attention. I was up late at night and early every morning with her and I gave her my Saturdays. I didn't mind because she was my little sister. I had another little sister who I took care of too, but she was older so I would just make sure that she did her homework and was looked after.

I wouldn't consider myself the primary caregiver because I still had both of my parents and they didn't neglect us. They may not have been there 24/7, but how many parents are? I could never say what it's like to be a teen taking on a full-time parental role, but this experience did make me more mature than most teens I know. I now know how to be responsible and how to take care of a baby. I know what it is like to care for a sick child and what it's like to be up in the middle of the night. Most importantly, I know how to depend on myself. How many teens can say that?

Discussion Questions:

1. What are some of the reasons teens take on adult responsibilities?

2. Who should neglected teens talk to?

3. What would you do if you thought a friend was being abused at home?

4. What situation is better: your parents being unemployed or overworked?

5. How do teens' home lives affect their social lives and performance in school?

Chapter 8

Teen Parenting

How can a teen mom be helped?

Today, teen pregnancy is a hot topic. Many people have a general view on it. We wanted to go about it a whole different way and shine the issue in a different light. We interviewed the people who are left in the shadows when it comes to this issue: the mother, the pastor, and the siblings. While trying to express the other side of the story, our own perspectives have changed. We learned that the issue of teen pregnancy is not as simple as it is stereotypically portrayed to be. We learned that even though there is a huge impact on the teen mother and her peers, her life and future aren't put on hold. Teens today have the ability to express themselves and make decisions.

Astrid Espaillat

By helping her to support her baby.

Would you have ever thought that your own friend could become pregnant? Would you help her if you were the only person she could trust? How? The way I would help my friend would be to give her my support, help her get through the pregnancy, and help her take care of the baby. Nowadays, there are a lot of teen girls becoming pregnant. Some, surprisingly, want to be pregnant and some do not. When my friend, Jessica, told me that she might be pregnant, she also told me that she didn't know what to do. I was very surprised when she told me, and my initial response was that I didn't know what to do either. I kept thinking about what her mom would think if she knew. Jessica's mom would be furious and maybe even kick her out of the house. I tried to reassure Jessica, but I could only rely upon my common sense because I have no personal experience with teen pregnancy. If she really was pregnant, the father of the baby would have to let Jessica stay at his house and I hoped his mother would accept her. The father had agreed to take care of the baby with her, but she also had issues with him.

Well, it just so happened that Jessica wasn't pregnant after all. She was relieved and so was I because I didn't want her to be kicked out of her house. Jessica's mother would have reacted so strongly if Jessica had been pregnant because she, herself, had been a teen mother and knows how difficult it can be. Her mother didn't want Jessica to go through what she went through when she was a teen. She had Jessica at the age of 17 and married Jessica's father, who was then 28, when she was 19. It is really hard to support a child on your own because you have to have enough money to feed it, put clothes on it, pay insurance and hospital bills, and also have money for yourself. You can look at all these adults who have kids and think that maybe you can have a child, but think twice. Having one is easy. Supporting it is the real challenge. There is always a time for everything, as my own mother told me, and that time flies. It's worth the wait because you'll have an easier and more fulfilling experience.

It was difficult for Jessica, not only because she might have been pregnant, but also because she didn't want to let her mother and father down. If she was pregnant, I would have helped Jessica by supporting her with the pregnancy, telling her to ask her mother some questions, and by being there if anything came up. Her life would have been extremely stressful because she also had family issues. Her parents were divorcing and she had to move.

Having a child isn't always a bad idea, but you have to be there every hour and every day just so that it can grow up into a healthy human being. Your life changes as soon as you find out that you're pregnant. "If you were a teen who just found out that you were pregnant, what would you do in a situation like that?" I asked Jessica. "[I'd wonder], how are we going to take care of it? Support it?" said Jessica. As a best friend, I told her that if she were pregnant, I would also help her. She knew that I would be there for her no matter what decisions she made.

SUGGESTED RESOURCES

Books:

The Color Purple by Alice Walker

The First Part Last by Angela Johnson

Teen Fathers Today by Ted Gottfried

Facing Teenage Pregnancy: A Handbook For The Pregnant Teen by Patricia Roles

Movies:

Fifteen and Pregnant

Life is Not a Fairy Tale

For Keeps

Momma 13

Just Another Girl on the IRT (R)

The Color Purple (R)

Music:

"Pain in My Life" by Saigon

"Runaway Love" by Ludacris

"Papa Don't Preach" by Madonna

"Baby Momma" by Fantasia

Carolyn Omar

Give teen mothers something to make their task a little easier.

Teen pregnancy has reared its ugly head in the crossroads of many unprepared teens' lives, including two of my family members. My eldest sister had her first born, who is now 21 years old, on the day before her 18th birthday. My sister's second born just had her first child, also at the age of 17. My sister and my niece are prime examples of teens who get a responsibility they aren't yet ready for, with high school graduation and college right around the corner.

"When people saw my face and my swollen belly following, they tended to turn their noses up," said my niece, an 18-year-old mother of a beautiful 10-month-old little girl, when I asked her if people treated her differently because she was pregnant. She told me that she never, ever would regret having her daughter; she may just regret what she did to get her. "No matter what people may think, my baby girl was not a mistake—just a really, really huge surprise—and she will never be unloved." Watching my niece, Nikki, I saw that having a baby is a huge amount of work that no teen can ever be prepared for. You have to make huge changes just to get your life halfway back to normal. "Meriah will always be loved but I won't lie. She can be four-handfulls and I only have two, so you can easily see were the trouble lies." My niece had to give up so much so fast. She couldn't go to prom and had to get a job. She basically had to leave her comfort zone and try to make a new one.

"Yeah, I thought about abortion and almost went through with it but then I thought about my mother and how, sometimes, it might have been. But we got through it. And, overall, we came out as fairly normal children. If she could do it, then maybe I could do it too. Also, I didn't have the funds so it was totally out of the question anyway." I know on TV they make wardrobe changes look glamorous, but watching my niece change her clothes every hour on the hour because the milk doesn't agree with Meriah's stomach is not at all glamorous. Not in the least!

I think there should be something better than food stamps or a little free formula for teen mothers because Pampers just don't buy themselves. When you're fresh out of high school and trying to go to college, without a scholarship and with a little job that pays $5.75 an hour, it gets harder to get simple things like Pampers, wipes, shoes and clothing for your baby. "Yeah, it would be much better if I had just a little bit more help," my niece told me. Teen mothers really do need a government-funded program made especially for them because I, personally, have seen more teens with children than adults with children. Also, when you're a teen, there aren't many job options out there that pay enough for you to take care of yourself, let alone another person.

My niece has opened my eyes wide to the good, the bad, and the ugly of teen pregnancy. "If you can prevent it, then prevent it with all of your might. But if it happens, cherish every day of it." The sin is in the act, not in the outcome. Teen pregnancy is just as big of an epidemic as small pox or the plague used to be. So, just as you do for diseases, you must protect yourself at all costs to prevent it or be willing to take on all the life changes that come with it. Like my niece has done.

Mariela Reyes

Invest in programs that support teen mothers.

Two-hundred seventy-three. The number astonished me when I read it in an article. Not because that was the death toll somewhere in Israel or the amount of hours children spend eating junk food, but because it was the millions of dollars spent on programs promoting abstinence in just one year. The total has come to one billion dollars since 1998. As I read more, I just sat in awe and disbelief. They are approaching the problem the wrong way. What happens to the teens who aren't able to get to these programs? Who are financially challenged or who live in a hostile environment? I noticed that the real situation is being overlooked; that is, helping teen mothers get through this harrowing experience.

To find out more, I spoke to Christine Dunburg, a sex specialist who works at a school-based clinic. It turns out that about one in three teenagers become pregnant before the age 20. 860,000 teen girls become pregnant and 425,000 give birth. A quarter of those girls are more likely to conceive again within three years after their first baby. On top of that, there are twice as many teenage pregnancies in the United States each year than in England or Canada and eight times more than in Japan. And, as James J. Fitzgibbon of Healthline.com explains, "Overall problems related to teen pregnancies cost taxpayers an estimated $7 billion per year." Where is all that money going? According to the Sexuality Information and Education Council of the USA (SIECUS), the government has spent a lot of those costly funds on programs to promote abstinence only until marriage.

"Never has so much money been spent in so many states with so little effect," reports James Wagoner, President of Advocates for Youth, in a criticism of abstinence-only programs. The government still continues with these programs despite little evidence that such programs reduce teen sexual risk behavior. Programs designed to prevent conception need to address other factors as well. All teen mothers have a different story and come from different backgrounds. Unless programs are going to take their time with every girl, they shouldn't have such a broad perspective towards the issue. It is unrealistic how much effort the government puts towards prevention when other factors such as environment, behavior, and development are ignored.

So what should the government focus on? Well, for one, let's focus on the teenage mothers who are currently struggling and trying to cope with this new way of life. Let's help the teenage girls and ease the heavy burdens they now carry by addressing all the factors affecting them. Christine Dunburg says that many young girls give up on school, thinking that it's no longer needed. We need to ensure that these girls attend school because, even though it's difficult, school is critical to the future of the mother and her child. The services and programs that should be funded, as much as the abstinence only ones, ought to be the programs that help the financially challenged. We should emphasize programs for young mothers that include parenting classes and day care, as well as programs that offer supportive services that will promote health and prevent child abuse and neglect.

In the end, it's a teen's personal choice whether or not to have sex. As the years go by, teens are becoming more self-governed in their decisions and are aware of the risks. The problem of teen pregnancy still persists but are abstinence programs the solution? Or would you rather not let taxpayers' money be wasted and, instead, invest in the real programs that do produce a solution. Ultimately, all I can say is be safe.

Sources

Fitzgibbon, James J. "Teenage Pregnancy". Healthline. 2002. Healthline Networks. 30 March 2007. http://www.healthline.com/galecontent/teenage-pregnancy

Wagoner, James. "New Report Details Money Wasted on Ineffective

Abstinence-Only-Until-Marriage Programs". Advocates for Youth. 22 June 2004. Advocates for Youth. http://www.advocatesforyouth.org/NEWS/PRESS/2004/062204.htm

Wikipedia. " How can a teen mother be helped" Wikipedia. 2006. Wikipedia Foundation Inc. http://en.wikipedia.org/wiki/Teenage_pregnancy

Discussion Questions:

1. Do you think teens are getting pregnant more often these days? If so, why?
2. Do you think that abstinence is the best way to prevent pregnancy? If so, why? If not, what do you think is a better way?
3. How can teens cope with being pregnant?
4. Do you believe parental involvement is important when a teen has a baby?
5. What steps can you take to help a teen mother who doesn't have any experience or enough money to support a child?

How does having a child in your teens affect your future?

Many teens get pregnant each year. Are we becoming careless? Is teen pregnancy now fashionable? What support networks are there for pregnant teens? These are important questions that need to be discussed. We chose this topic so that our readers could better understand how having a baby in one's teens can affect your future.

Jeffrey Raymond

You have to make decisions that you might not be ready for.

It all started four years ago when I asked my mother about my father and about how she was a teen mother. When I interviewed her, I got the rest of the truth.

Ms. Helena Cruz, my mother, was an 18-year-old when she was pregnant with her first child. She had just finished high school, and the father of the child was nowhere to be found. In June 2003, the best person in my mother's life—my grandmother—passed away. It was a very hard year. In the middle of July that same year, I was in my house when the buzzer rang downstairs. My mother thought it was a package and brought me downstairs with her. There was a man with a girl. I didn't recognize them at all. My mom seemed to know the man and went off to talk with him while she sent me to talk with the girl. The girl was really nice. Later that day, my mother told me that the man was my father and the girl was my sister. It was one of the hardest days to live with, especially after the passing of my grandmother. Up until then, I thought my father was my brother's father. Looks like I was wrong.

When I interviewed my mother, I asked her how she felt about teen pregnancy before and after she became a teen mother herself. She replied by saying that she felt it would be a very hard thing to do, and that teens shouldn't even be having sex because they are too young and immature. She continued by saying that no one really talked to her about pregnancy when she was young, let alone teen pregnancy. She said that after she became pregnant, she felt the same way. She didn't know what to do so she lived with my grandmother for a while. I also asked her what she would have done differently if she had the choice. She told me that she would have waited until she finished her childhood and her education.

"I think teens should be educated about having unprotected sex," my mother said. I agree with her because teens these days obsess about sex so much. If teens don't have proper education and knowledge, they can end up with an unexpected pregnancy. I wouldn't want other teens to feel the way I did that day I met my father and was like, "What the hell?" It was confusing because the man who is my father is not the one who treats me as his son. I actually think that if my father was around when my mother needed him, it would have been a better situation for all of us. I wouldn't be so confused about everything and he would have actually been there for his son.

That's why teens should think twice before having unprotected sex. You don't know whether you are getting yourself into something great or something bad. As my mother said, "Wait until you are old enough to make good decisions."

SUGGESTED RESOURCES

Books:

The Girl with a Baby by Sylvia Olsen

Annie's Baby edited by Beatrice Sparks

The First Part Last by Angela Johnson

What Kind of Love? by Shelia Cole

Movies:

Lean on Me

Coach Carter

Fifteen and Pregnant

Too Young to be a Dad

Websites:

Teenpregnancy.org

Helponteenparenting.com

PlannedParenthood.org

Teenwire.com

Yamilka Rivas

It makes your dreams even more precious.

Where I live, there are a lot of girls my age getting pregnant. I have to admit that I think some teens actually want to get pregnant. I think they do it to get their boyfriends to stay with them or love them more. I also think some girls do it to grow up faster and to have a small human being to love them. But the reality is that it often makes relationships more difficult and it's harder to find yourself or reach your dreams when you're raising a small child. I was brought up to believe that if I had a baby at a

young age, I would have to leave my home and live with my boyfriend. My family made it clear that they would not put up with any teen pregnancy in the family. Consequently, I simply thought that being a teen mother would ruin my life. Then my cousin made me see things a little differently.

Julia is smart, hardworking, well spoken, and loves to spend time with her friends. She is also a 17-year-old teen mother. When she was 15 years old, we heard from her mother that, for no apparent reason, she had moved out of home. But I suspected that something was up. Was she just having a fight with her parents? Did she want to live with her boyfriend and be more independent?

Finally, she cleared things up for me. She was moving in with her boyfriend, but it was because she was pregnant and her father had kicked her out. I was scared for her and, honestly, I was worried about her future. She is such a smart girl and I didn't want her to throw it all away. When she was about four months pregnant, she moved back in with her parents because they had realized their mistake and decided to support her. Becoming pregnant at the age of fifteen was not "a dream come true", but now she is on a mission to show the world that it is not a complete nightmare. Having a child at the age of 16, Julia knew that nobody believed she would succeed but she has worked hard to prove everyone wrong. She continued to attend school up until she was seven and a half months pregnant. From then on, she went to school weekly to do the work she needed to complete high school. After giving birth, she took her Regents Exams and is now expected to graduate on time in June. She still plans on attending college to fulfill the dreams she had before she got pregnant. The only difference now is that she'll have someone with her along the way.

Julia is a positive example of teen motherhood because she does not let anything or anyone get between her and her dreams. She stayed in school and still found time to raise her son. She does not regret having a child; she actually thinks it is a blessing. She believes that she wouldn't be succeeding as much without him because he is her inspiration. "Having a child at a young age is not the end of your life. You can still follow your hopes and dreams. It is the beginning of a new chapter of your life."

Looking at Julia, I no longer believe teen pregnancy is the end of the world. The mother has to be willing to work extra hard but, if she does, she can still reach her dreams. I don't plan on getting pregnant until I finish high school, but it's good to know that opportunities are available for teen mothers if they are willing to put in the effort.

Sasha Strongbo

It makes life more challenging.

Many teens who have unprotected sex do not think about how becoming pregnant will affect their futures. My mom has such high hopes for me and having a baby while I am still a baby myself would destroy her. I have seen my sister struggle as a teen mother and it has been very hard for her. She had to take care of her baby while she, herself, was still growing up. When you are a teenager, you are still developing and growing into the person you will be as an adult. The fact that adults find it hard to raise children gives you an idea of how hard it can be for a teenager. If you do have a baby, however, it's not like it's the end of the world. In some cases, it might be a good thing—your baby could be the next Donald Trump! But even so, it's still better to wait.

Teens could be taking on huge responsibilities by not protecting themselves during sex. I interviewed a former teen parenting counselor to gain her perspective on teen pregnancy. She said that teens get pregnant for many different reasons, which ultimately depended on the person and their situation. Some teenage girls get pregnant because they are looking for love and attention from their sexual partners. They think that if they have a kid, the boy will never leave them. Others are just immature.

I have a friend who is currently a sophomore at Humanities High School. She gave birth to her baby, Amanda, at the age of 14. Her life is very different now that she has a daughter: no more late night parties, going out, or chilling with her friends. All her spare time is now spent with her daughter. It has been very difficult for her. When Amanda came into this world, my friend would ask herself, "Why is she crying?" "Do I feed her now?" "She's sick, what do I do?" These questions ran through her mind like crazy. But despite the stress and responsibility, my friend, who grew up in foster care so never really had anyone to care for her, has also found great comfort in Amanda. This important responsibility has also inspired my friend to seek out a better life for herself and her daughter. She plans on moving to Virginia and starting fresh.

I think that teens should take care of themselves more. I know many teens who don't use protection and are not on birth control and it could affect their future. Their lives will never be the same. Teens are not taking teen pregnancy seriously. They should take care of themselves now rather than complaining later that they don't know what to do, when the whole dilemma could have been prevented. I have a friend who would always use protection until the guy she was seeing said that he was allergic to condoms. They haven't used one ever since. I guess they think it feels better, but I say it's better to be safe. When teens first start having sex without protection, they may like the feeling and they might say to themselves, "Hey, I didn't get pregnant the first time so if we use the pull-out method I will never get pregnant." Ladies and gentlemen, don't fool yourselves! This is not a safe or foolproof method of contraception. Not only can you get pregnant but you could contract an STD, some of which are incurable. Clinics, your doctor or other health professionals can provide you with birth control without your parents knowing. So why are so many teens getting pregnant?

Discussion Questions:

1. Why do you think teen pregnancy is so controversial?
2. Why do you think some teens are uneducated about safe sex?
3. Do you feel we need more sex education in high school today? Why or why not?
4. What are the consequences of teen pregnancy?
5. Do you think most teen pregnancies are intended? Why or why not?
6. What drives teens to have unsafe sex?
7. What is the appropriate age for sex education?

Chapter 9

Violent Situations

What should we know about sexual abuse?

Have you ever been abused and needed someone to talk to? Well, in this chapter we are going over the experiences and choices of the witnesses of abuse and the people that have been abused. There are many stories that need to be heard. These stories, facts, and opinions will help you decide the choices you could make. This topic is touchy among teenagers because it makes us feel uncomfortable, but that's exactly why we should talk about it. In this section we hope to have many people learn from one another. We want to end this cycle.

Gervoni Brockington

Everyone should be aware of how they present themselves and the situations they are in.

For many years I have always wondered what would provoke a human being to force another person to have sex with him or her. When I was just 13, I sure did find out. It was my first teenage party and I picked out my cute denim mini-skirt outfit. As I walked out of my room and said bye to my mother, she pulled me back and told me, "Take that crap off! You don't wear no skirt to no party!" So I went back into my room and put on jeans, but I told my friend that I was going to put the skirt in her bag.

In the staircase, I changed into my skirt and left my jeans there. When we arrived at the club where all the cool teenagers hung out and partied, every boy was trying to grab me and talk to me as we walked up to the doors. I got a little scared, but I knew my friends would have my back. Then suddenly this tall, brown-skinned boy with braids grabbed me and I smiled and my friends pulled me along. As I made my way through the crowd I started dancing. A friend I hadn't seen all week was there and we greeted each other. He walked towards us and the boy introduced himself, but my friend wasn't impressed. She had a look on her face of disgust and hatred. When I asked her what was wrong, she replied, "He raped me!" I asked her what she meant. "The last party that was here he took me into the bathroom and raped me."

As her friend I thought I had to do something about it, and I eventually did. But then and there, I asked him to talk to me outside. As we began to talk, he started touching my leg and I told him, "Hey, I'm not that kind of girl!" He explained that he could respect that, so I asked if he had attacked my friend.

"She wanted it and she knew she did," he responded. I don't know why she keeps telling people this." I sat down and asked how she was asking for it.

"Because she was wearing the same thing you're wearing tonight—that mini skirt showing off you curves and—damn girl!—I must say you sexy!" I thanked him but explained that she had said no, hadn't she?

He paused for several moments and said, "Look, let's talk about better things, like me and you."

"I don't think so, Sweetheart. You aren't my type," I replied. After that night I haven't seen nor heard from him. From that night on I learned a lot about how your appearance can encourage a physical attack, even if that's not what you are looking for.

I interviewed a person I thought had the most insight on this topic. Her name is Deborah and she has been a registered nurse at Lincoln hospital in The Bronx, New York, for over ten years and has worked with many victims of rape. "During my stay there I have worked with over 250 rape victims. While working with rape victims, I learned more and more everyday," she said when we spoke. I asked her if she found any clues that lead her to believe that the attackers go after a specific girl. She responded, "In my own experience, I have noticed that there are two types of females that attackers might go after. They might be the popular girls—the ones who are the glamorous types. Or they might be the silent girls who keep to themselves."

Deborah was once raped herself. "I always thought it was my fault. As I grew older and talked about it, it kind of eased the pain I had. I remember one day I was working late and this girl came into the emergency room and she was crying, but to herself. I asked what was wrong and she replied that her boyfriend just took her virginity without asking. This girl told me she was only 13 and her boyfriend was 16. Then she asked me, 'Is this my fault? If it is, please, can you tell me?'"

SUGGESTED RESOURCES

Books:

Speak by Laurie Halse Anderson

The Color Purple by Alice Walker

Push by Sapphire

The Bluest Eye by Toni Morrison

The Truth About Rape by Teresa M. Laver

I Never Called It Rape: The Ms. Report on Recognizing, Fighting, and Surviving Date and Acquaintance Rape by Robin Warshaw

Invisible Girls: The Truth About Sexual Abuse by Patti Feuereisen and Carolyn Pincus

Movies:

Antwone Fisher

The Color Purple (R)

Websites:

www.nextstepcounseling.org

www.essentialchange.org

www.ascasupport.org

www.ra-info.org

Deborah stood there stone cold and paused for a second to think if, when she was in this girl's shoes, it was her fault. "'No, it's not your fault, darling,' I told her. "'It's no one's fault but that pig that raped you. He took what belongs to you.' Then I took her to do the procedure for rape victims. Before I sent her home, I told her that if she wanted to press charges, I would be there. To me, that day changed me, not for the worse, but for the better."

According to The National Sexual Assault Online Hotline, every two-and-a-half minutes, somewhere in America, someone is getting sexually attacked. One in six American women and one in 33 men are victims of sexual assault. From 2005 to 2006, 200 to780 incidences of rape were reported. About 44% of rape victims are under the age of 18 and 80% are under 30.

These statistics, Deborah's story, and my own experiences have taught me that we all need to be careful and guard ourselves against those who want to hurt us. All I have to say is if you have been raped or abused by someone you know or someone you don't know, it isn't your fault. Seek help. Go to someone who you trust and ask them to help you make the next step and do something about it.

Source

"Statistics" 2005. The National Sexual Assault Online Hotline. The Rape, Abuse, and Incest National Network (RAINN). February 22, 2007. (http://www.rainn.org/statistics/index.html)

Jessica Harvey

You can protect yourself by never letting your guard down.

I see on the news that there are many women being attacked, but mainly for the same reasons. They let their guards down, come out at a certain time in the night, or leave a party intoxicated. Are those the real reasons? I've noticed that the people most often attacked are those who are defenseless and alone and wearing suggestive outfits. I wanted to know if there was anything that we could do to make sure we are protected whenver we are out. Should we wear sweatpants or cut our hair? What steps should we take to keep ourselves safe?

I got in contact with Ms. Frolayne who works with a group called San Francisco Women Against Rape (S.F.W.A.R). It's an organization that works with individuals who need help and healing because of their struggle to end sexual violence within their lives. She allowed me to interview her and get my questions answered. I asked Ms Frolayne if she ever had her own experience with a predator. She responded, "People are drawn to this work for personal reasons, whether personally having had experiences with a sexual predator, or realizing that every member of society is touched by sexual violence in some way, shape, or form." I see that some reasons are more personal and troublesome than others, but everyone has their reasons. She told me that rape is defined as a forced sexual act. She explained that "force," means using pressure, fear, intimidation, and manipulation to physically, emotionally, or psychologically make someone have sex. Sometimes drugs or alcohol are also used in sexual violence. When she told me these things it made me think, "Wow! There are a lot of ways people can be attacked and sometimes not even know it."

Next I asked, "Are there any ways of knowing if you're around a predator?" She responded, "It's about following your instincts. There's no stereotype of a predator. It could be a trusted friend, loved one, family member, or some one you met ten minutes ago. You can be attacked by someone and not even know the reason. Just follow your gut." At any moment you feel uncomfortable around anyone just walk away or make sure you're surrounded by people you do trust. "At no point does anyone deserve to be violated, even after a kiss or a fondle, or after entering the person's home or room, or wearing a short skirt. You have the right to say no at any point. Even if you were in an intimate situation, that person should respect your wishes." Basically, understand that no matter what no one should violate anyone's rights, no matter the situation. At any time you wind up being in a situation like this, please, I beg you, don't be afraid to speak up and don't keep it locked up inside of you. Let your feelings out.

Ms. Frolayne's advice has influenced my own thoughts on attacks in many ways. The key is to be aware of my surroundings and myself. We can't just stay in the house all day, but we can surround ourselves with safe people. We know those neighborhoods we should avoid at certain times. If we are attacked, we should make a lot of noise and get people's attention. If an attack occurs, we should go to the hospital or the police immediately. A woman who wears sexy clothing is not "asking" to be raped or attacked, so she is not to blame. But we should be careful with the messages we might be sending without knowing it.

Yahel Madera

"Abusers" might not even be aware of what they are doing.

"What do you think provokes a predator?" was the question I asked a friend who had an experience with this topic. The first thing she said was, "We minors look a little more mature than our actual age." And I really agree with her response, but that's something we can't control—how and when our bodies develop. But the way we take advantage of our bodies can be controlled.

Like I said, a lot of minors tend to look older than their age and some men fall for it. 23-year-old Tony was one of those men. He met a young female named Katlyn who looked like she was in her early 20s, claimed to be 18, but was really only 15. So he believed her and they started dating. Four months into their relationship they began to engage in sex, and they fell deeply in love.

One day Katlyn mentioned that her mother wanted to meet him. Tony was willing, but for some reason Katlyn had a problem with that. They soon got into a little argument and Katlyn started crying. She finally gave in and decided to introduce them, but not before saying something that made him feel awkward: "If you find out something about me, would you leave me?" she asked. He replied, "It depends." When they got to the house, her mother went ballistic. She began staring at Tony with her mouth open and kicked him out of the house.

The next day, when he was in his apartment, the police came knocking on his door and told him that he was being arrested for statutory rape. He didn't understand anything that was going on. He began to understand when he got to court and the judge announced the case and said Katlyn's and her mother's names loud and clear. He was angry and in shock, not only because of her age, but also about what Katlyn's mother had her saying about him. He knew her mother got her to say those things because her mother would just keep adding something onto the end of each and every sentence that Katlyn said out of her mouth. The only thing she was able to get out was that it was a lie. But the judge couldn't see past it. To him, she was crying because of guilt. She didn't look at him even once after the judge

announced the charges. Tony was put in jail for a year and a half.

He was three months out of jail when he saw her again. When he did, he heard big news. She came to his apartment with a baby girl and claimed that it was his. He knew it was because the baby had his exact face. Every once in a while she would bring his baby to see him. They began to see that they did love each other very much and Katlyn's mother couldn't do anything about it because she had his baby and he would be in both Katlyn's and his daughter's lives. At this point, they are still together.

I can understand what Tony's issue was because, when you fall in love, that's all you can think of. But people under the age of 18 are minors for a reason–they are too young to make decisions that could affect the rest of their lives. Tony and Katlyn learned this the hard way.

Danica Young

Family members can be the abusers.

Many victims of sexual assault are afraid that they will be blamed or not believed by others. One of the reasons why it's hard for girls, boys, and women to report rape is because it often happens between two people who know each other, making it hard for the person to prove it ever happened.

"Sexual abuse takes away the victims' power and control," said one of my friends, a victim. "When I was raped by my uncle, I blamed myself for the abuse because I thought to myself that I acted too grown up for my age or the way I dressed. But that couldn't be. I was too young. I tried every excuse in the book to blame myself but now I realize it wasn't my fault," she says. "See, it started out by me being molested at the age of six. I didn't know what he was doing till my aunt started talking about sex around me. Then I realized he was doing something wrong," she said.

"People don't understand what it's like to be raped or molested. The only way they can understand it is if they were a victim," she added. "I kept my mouth shut because I wanted it to go away and, for awhile, it did. It comes back when I think about it. When you don't, everything seems okay. I forgot about it and, for years, my memory was a blank, but I never will forget his face. God knows I would take this to my grave," she said, knowing it will be with her for life. I was sitting there just looking at her because I felt for her and understood what she meant when she said that you couldn't understand their pain unless it had happened to you. The things she said made me feel like I wanted to cry. Getting molested at the age of six is horrible; I can imagine her being scared, not knowing what's going on, and lost in her mind. Who would do such a thing?

I checked out the online magazine, "Advocates for Youth", and found some disturbing statistics on family sexual abuse. First of all, it is important to know that sexual abuse can happen to anyone anywhere. According to the website, "Sexual abuse occurs in rural, urban and suburban areas and among all ethnic, racial and socioeconomic groups." This problem affects us all. The Department of Justice states, "Half of the women who reported rapes in 1992 were under the age of 18." Also, a study conducted in three states found, "96 percent of reported rape survivors under age 12 knew the attacker." But these statistics may not be accurate, because many victims are afraid to come forward and say they were abused. Some are ashamed of what happened to them or have repressed memories or do not admit that it was real abuse because it was by a family member.

Whenever a person is forced to have sex with someone, a rape has occurred. It doesn't matter if someone rapes a stranger, a friend, a girlfriend, or his own wife. Even if you had sex with that someone before, no one has the right to force you to be intimate, including a family member. There are so many people ready to help victims of abuse. Guidance counselors and teachers will know what to do. And there are many free hotlines that can give professional advice—and you won't have to give your name. You just need to know you're not alone and you don't have to deal with it by yourself.

Source

Finn, Susan K. (editor). "Child Sex Abuse I: Knowing the Facts". Advocates for Youth. January 1995. Advocates for Youth. 30 March 2007. http://www.advocatesforyouth.org/PUBLICATIONS/factsheet/fsabuse1.htm

Discussion Questions:

1. Who or what in your life made you feel safe?
2. Describe a time when you didn't feel safe.
3. What is a predator? Where does the drive to attack someone come from?
4. Describe some places and situations that could put you in danger.
5. What do you think schools should do to keep students safe from predators?

Why do people abuse others?

Abuse is a very uncomfortable subject. People who have experienced abuse might be afraid to talk about it. It's especially difficult when it's happening in your own home or with your boyfriend or girlfriend. If someone is too afraid to talk then you might have to give them some space, but it is important to eventually listen to their stories and support them.

Chayil Anglero

Abusers feel like they need to be in control of the situation, and sometimes they were abused themselves.

I have always thought people abuse others for many reasons. The obvious ones to me are problems at home or ignorance about other ways to deal with problems. Maybe these people need to take out their problems on those who are close to them, like a girlfriend or boyfriend. Maybe hurting others lets abusers feel a little bit better or more confident, like a bully who might have been abused him or herself. After a while, they get addicted to abusing other people because it makes them feel like they're on top instead of being on the bottom like they were when they were abused.

I have a friend who likes to beat up on his girlfriend. I told a guidance counselor about it because it was a bad situation. Every time I saw him beat her up, I asked him why he was doing it. I warned him that she was going to call the cops on him. He always answered by saying that she doesn't think and doesn't listen to him. I thought to myself, "That's the reason? That's dumb." He often tells me terrible things about her, but I do not understand why it leads him to use physical violence to solve problems. It's true that sometimes men use violence to solve their problems, but to fight a girl is wrong and it doesn't prove a man is tough. It just proves he is crazy and weak because he doesn't use his brain to solve problems.

Another problem with my friend's situation is typical in movies where there's an abusive relationship: the girl keeps coming back for more. I wonder if some girls become used to it and think they can't live without this person who is hurting them. When I first saw it happen, I wish I had told my friend's girlfriend to get away from him or get a restraining order or do anything that would prevent him from ever seeing her again. Too bad we can't get these girls bodyguards.

I wanted to find out why people abuse those they love so I checked out the specifics about abuse on the website: www.metalhelp.net. According to the website, abusers are often victims of abuse themselves. As they get older, they often have the choice of being the victim or the abuser so sometimes they choose to be an abuser because they want to be in control of the relationship. Because the abusers were once the victims, they know many ways to abuse their partners. By choosing to have control, they have it in their minds that no one can hurt them and they can rule the other one in the relationship. That may be part of the reason why they keep going on without caring about others, as if they are the only ones in the world. You can help your friends seek therapy. The first step is to tell a guidance counselor, like I did.

Kevin Ong

People abuse others because they are depressed, angry, and/or intoxicated.

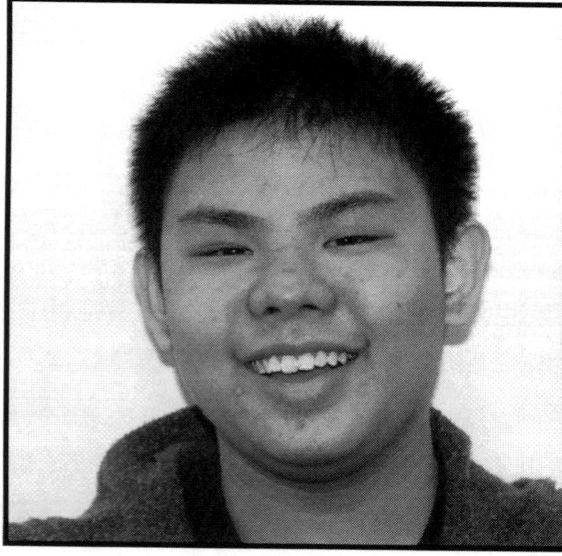

What is abuse? Is it using words or violence? For me, abuse is hurting others physically or mentally, and maybe using weapons for revenge. Before I researched this topic, I assumed a person abuses others because he or she was abused in the past. Movies and television made me believe that it stems from bad family relationships, excessive drinking or drug use. But after interviewing a victim on this subject, I found out it's more complicated than the way it is portrayed in the media.

Jane, my mother's friend, lives in Staten Island and works in a bakery. She's a caring person and brings fresh-baked goods whenever she visits. I always look forward to seeing her. It's sad that such a positive person had to deal with an abusive husband. Before they divorced, Jane's alcoholic husband would beat her after drinking too much. He would beat her for a few days when on a drinking binge and then she would finally call 911 and tell her friends. She would come over to our house to discuss the situation with my mother. From my room, I could hear the details of the latest fight. It sounded like it all came from his drinking problem. Later my mother would shake her head and tell us how stupid she thought he was for hitting Jane.

When I asked Jane directly on the phone about the situation, her perspective on this issue was that abuse is wrong and it usually comes about because of the abuser's anger or frustration toward a bigger situation. She was very open about discussing the abuse and said matter-of-factly, "I hate my husband and now my kids hate him, too. When he calls now, they just hang up the phone." She assured me that life is better now that he's out of the picture and they're in a different home. The children have more freedom and can do more things. She's not afraid he'll ever go after her because he has found someone new.

I think the "bigger situation" that Jane thinks causes someone to abuse others is whether someone has a fierce personality or if they are deeply depressed about something in their life. I'm still not sure what this man was depressed about, but as soon as I heard that he had abused her, I was more concerned about his kids, who are my friends, than about his mind.

SUGGESTED RESOURCES

Books:

A Child Called "It": One Child's Courage to Survive by Dave Pelzer

Bastard Out of Carolina by Dorothy Allison

The Breakable Vow by Kathryn Ann Clarke

Movies:

Holiday Heart (R)

Enough

Websites:

www.kidshealth.org/teen/your_mind/families/family_abuse.html

http://en.wikipedia.org/

Discussion Questions:

1. What is your definition of abuse?
2. What are some of the reasons people abuse others?
3. Has abuse ever touched your life? How did this experience affect you?
4. If you were abused, whom could you ask for help?
5. Does abuse affect your future? Can you recover from it?

Chapter 10

Tough Questions

Why do people discriminate?

What is discrimination? Discrimination is a strong word for what happens when people judge others based upon their clothes, age, sexual orientation, gender, language, race, etc. It is important to talk about discrimination because almost everyone has experienced it; even victims of discrimination can discriminate against others. Sometimes people discriminate without knowing how it will affect others' lives, not realizing that some people who are victims of it might mistreat themselves because of it. People who discriminate tend to break others down by hurting and teasing them, which can contribute to making someone feel depressed or nervous.

Jonathan Charles

People discriminate because they hate what they fear.

What is discrimination? To me, discrimination is the act of treating someone differently because of his or her race, gender, sexual orientation, or class. My own upbringing has taught me that people discriminate because they are afraid of what is different, and they want to separate themselves from people who are looked down upon in the community. Take kids in school. In that environment, you basically have to have the right attire. If not, then you might experience constant teasing from kids saying that you're poor. Personal choices also play a role in whether you'll be treated wrongfully or not. For example, if you're homosexual, many of your peers might shun you. They might test you, tease you and push your buttons until you can't take it anymore. And if you don't find help somewhere, you might start hating and mistreating yourself. It's a terrible

cycle that is started with hatred.

"I think that it's a sickness, an illness. To be that way, it must have been due to brainwashing." That's what my mother said when I asked her what she thinks about discrimination. Discrimination has been around forever but the type of discrimination the public is perhaps most familiar with is discrimination among the races, or racism. Many people have either experienced it firsthand or have used it against other people. "It's not normal," my mother said. "When a child is born, he or she doesn't go looking for an enemy; they go looking for a companion, a friend to play with. When they start hating others because of their skin color, it means that they were taught to hate someone because of their race."

My mother was born in Farragut in Brooklyn, New York. Back then, Farragut was known as the projects but it wasn't as bad as it is now. In her neighborhood, she grew up around a mixture of races like blacks, Jews, Italians, Puerto Ricans, and Irish. Since she lived by the Navy Yard, there were many other races that came from overseas. So, even though there was racism in New York, there was very little of it in her neighborhood.

Though she knew about discrimination, my mother was brought up with the Bible and believed that all men were created equal and that hatred is a sin. As a kid who spent her vacations in the South, my mother was taught how to identify discrimination by her parents. My grandparents gave my mother rules whenever they were there. You couldn't look a white man in the eye; you had to move out of their way when walking; you had to use separate fountains and sit in different areas. Whether you were a kid or an adult, if you wanted to be safe, you couldn't walk the streets by yourself or be out after dark.

Recently, I asked my mother if discrimination was an issue in her childhood. I already knew the answer but I wanted to get more information. She told me that it was bigger than I would've thought. During her years as a teen in high school, it got to the point where my mother had to join a civil rights revolution to fight for equality. There were only five schools that would let her attend while the other schools were strictly for whites. Blacks were also excluded from schools for special careers and the arts. Their education, however, wasn't the only thing suffering. Plenty of friends were hurt or killed in the hate crimes that were brought upon them. Sitting around and trying to work it out was a not a factor when the pursuer was bent on killing you.

"Racism made me a stronger person, a fighter, not only physically but mentally. It made me learn how to fight the system. For example, I had a former teacher and she was prejudiced. Even though I did well in her class, she failed me. So I switched homework papers with the top student who was white, fooling the teacher. The teacher gave me a zero and the other girl an A with my paper. That right there made me a stronger person and that's why I push my kids so hard by telling them to excel in school, learn their history, and mentally train themselves

SUGGESTED RESOURCES

Books:

To Kill a Mockingbird by Harper Lee

Night by Elie Wiesel

Roll of Thunder, Hear My Cry by Mildred D. Taylor

China Boy by Gus Lee

Jack by A. M. Homes

Issues on Trial: Racial Discrimination by Mitchell Young

Movies:

Crash (R)

The Color Purple (R)

Glory (R)

Philadelphia (R)

Swing Kids

Racism on Campus

Hotel Rwanda (R)

Rosewood (R)

GI Jane (R)

Songs:

"Minority Report" by Jay-Z

"Faces in the Hall" by Gym Class Heroes

for anything that comes their way because you never want to fall for someone else's ideas or beliefs. You want to have your own ideas, your own beliefs, and never want to fall into anyone else's trap."

Before my mother's interview, I knew how bad racism was but when you hear it from someone who is close to you, who has lived through it, that's when you open your eyes and pay more attention. It is not a great thing to experience being told that you can't play with someone because of their race or being on the receiving end of teasing, especially as a child. I, myself, have personally felt the pain of discrimination from a friend's parent. "It's a sickness," my mother said and I agree with that. Hating someone isn't hereditary. It is taught, and it is a dreadful thing that's not worth learning.

Justine Montalvo

People are not ready to accept each other's differences.

"Don't worry about what other people think because it's you living your life, no one else. You have to do what makes you happy, not to please everyone else," stated Jessica. Jessica knows what it is like to be discriminated against because she is gay. One person who discriminates against her is her mother. Jessica's mother calls her names and states that gay people are nasty and will get nowhere in life. She also refuses to let Jessica see the love of her life and even bribes her to become straight. It hurt Jessica the most when her mother tried to kick her out of the house because she did not want her homosexuality to rub off on her little sister.

Discrimination is a strong word. When someone is discriminated against, it can break them down. I had my own experience with discrimination when someone wanted to fight me just because I am Hispanic. I did not think it could ever happen to me, but it did. When I was in 7th grade, I got into a fight because some girl did not like me because of my race. This made me want to know why people discriminate against others.

I think discrimination of gay people is unique because a lot of people think it is immoral to accept them. Once I was walking and I saw some kids beat up another kid just because he was gay. Often I hear males talk about gays as if they don't have a problem with them, but once a gay guy casually touches them they will just hit or stop talking to them. I think they do this because they feel uncomfortable and they don't understand that you're born gay. You don't just "catch it" like a cold. Jessica believes that gay males are discriminated against worse than girls because most guys think that it's OK for a girl to be with a girl but not for a guy to be with a guy. They think it's wrong and not "normal." People handle situations differently. If I had to say one thing to people who are discriminated against, I would tell them to not worry about what other people say. You can't let them break you down. You have to be you, no matter what.

After speaking with Jessica, I found out that people's lives are not as perfect as they seem. "It's true what they say. You can't judge a book by its cover." So the next time you want to judge someone, try to understand where he or she is coming from and what they might have been through. Everyone needs support and loved ones by their side.

Elena Pena

People can discriminate when they don't really know someone.

Discrimination. A word I don't find myself connected to but am interested in. I have never discriminated against anyone and never would. I have never been discriminated against and would not want to be. My interest in this topic is to know how people feel and what they think about it. I would also like to know what type of things people have had to go through and how they got through them. I have a friend who was able to answer my questions because she has had an experience with discrimination.

Rosa is 15 years old. She's Mexican American and lives with her parents, three sisters and one brother in East Harlem in Manhattan. She doesn't work, but she's on her school's track and field team. She's currently in the 10th grade in a private Catholic high school in Brooklyn. Her old coach, Alfonso, gave her the hope and help she needed to receive a scholarship to enter the school she's in now. Quickly, though, she began to regret everything she had to go through to get that scholarship.

"At first, I felt happy and excited because I got accepted to my school. But as soon as I entered my class, everyone just stood staring at me. I was the only Hispanic there," said Rosa. In her school, 95% of kids are white, 4% are black, and 1% is Hispanic. "The first two and half months I went to school, nobody would talk to me. Then, in December, I started to practice with my track team. I made friends with three Latinas and three crazy white girls. When I went to practice, none of the girls of other backgrounds would talk to me or my friends. In the middle of February, this group of white girls started trouble with us. The conflict was about my three non-Hispanic friends who used to hang out with them. They told my friends that if they still hung out with us, they would never talk to them. The group of girls told them that we were worthless trash and that people from other backgrounds don't go together. My friends were shocked at what they were saying and told them that they were still going to hang out with us."

In February, there were a lot of important races and Rosa's coach always chose white people to run them. Rosa thought that it was because they were better than her but in practice she would always beat them. Finally, her coach noticed and told her she was good. When Rosa heard those words, she was glad. She was also happy because she was the third best on her whole team. "The Madison Square Garden race was coming up and I thought I was going to get chosen but I didn't. I asked my coach why and he said I needed a little more work. I stood in shock because he chose the girls that I always beat. This wasn't the only race he left me out of. I didn't run a race 'til June. I only ran then because one of the girls didn't come. In the beginning, I was furious because I was a replacement but at the end of the race I won first place overall. Everyone stood in amazement. I was glad because I qualified for the semi-finals."

Rosa went on to say, "The next day my coach called me to speak with him and told me that I wasn't going to run in the semi-finals. I asked him why and he said that another girl was going to run it. I told him that I was better than her and that it was because of me we got in the semi-finals. Then he told me the craziest thing. He said that white girls had greater strength than Hispanic girls. I got angry and told him that just because I'm Latina doesn't mean I can't run. I told him I was going to tell the principal and I ran out. I told the principal everything and he called my coach inside. My coach didn't deny anything and repeated the crazy thing again. The principal stood in shock and told him that he wanted the race times of every kid on the team. The principal saw that I was the third best and told him that I was as good as everyone else. The next day, the principal told me that he had fired my coach and wanted to talk to everybody on the team. He explained that we should all get along. He made all of us talk to each other. We started talking and learned things from one another. We found similarities and solved our differences. I noticed that my coach never made us talk to each other or work together and we didn't mind doing so."

Lonely, different, and sadness were the words Rosa used during the whole time I interviewed her. "They would make me feel low, down, like nothing," she explained, her face full of anger. She also said something that caught my attention, "I've never felt unwanted like I did on my team." Rosa says that she didn't know why they treated her like that in the beginning. They didn't even know her or who she was. They only looked her up and down and walked away. "I'm as equal as everyone else. Last year was the worst but that was the beginning. Now I'm doing well. I got to know new people and they got to know me. I get along with everyone in my team. I look out for them and they look out for me."

Even though my friend explained how she felt, I couldn't relate to her. She went through something I hope never to experience. This has influenced me by opening my eyes. Never judge someone by how they look or where they come from. Think of the person's feelings and treat them equally. Discrimination is when a person dislikes another person because of their color, gender, or religion. There's a saying, "If everyone is the same then it would be a boring world," so why hate someone because they like different things? This is America, the land of dreams and freedom. Why are people still acting like it's not? What is democracy? Is it having equal rights no matter what color, gender, or religion a person is? Why are some people taking these rights from others? Anyone can be discriminated against and that's bad because that is what we are teaching the future of America. We might come from different places but we all have the same dream.

Discussion Questions:

1. How can we talk to or help people who discriminate against others because of race, sex, gender etc.? What might we say to them?
2. How could your school or community improve the way people treat each other?
3. Why do you think teens discriminate against each other?
4. Have you ever discriminated against someone? If so, describe the incident.
5. How are people of different genders, races, etc. portrayed in the media? Do these portrayals encourage or discourage discrimination?

Why do people join gangs?

Are teens to blame for all the violence and gang activity that is occurring in society today? Everywhere we turn, we hear about someone who just got shot and killed. To help us better understand this issue, we chose some people to interview on gang violence—one who was a teen gang member himself. These people all had their own perspectives on gangs and gang violence. It was a challenge to make them comfortable enough to talk deeply about their experiences and opinions on gang activities. The interviewees gave us great feedback on why they think people choose to be part of a gang. After the interviews, we learned more about gangs and considered the varying perspectives of the people we interviewed.

Leslyn Black

People join gangs to feel supported and accepted.

"Stand clear of the closing doors." I had just entered the subway, which was usually packed, and was forced to stand between two people with little space. Suddenly, a group of about ten boys sprinted onto the train, shoving everyone out of the way. As the train came to a halt, everyone quickly dashed out. I wanted to know why. Was that the last stop on the train? I glanced in front of me and spotted two boys fighting. One boy fell and banged his head against the seat. When he fell, two other boys joined in and started stomping on his face. Everyone who remained on the train minded their own business, except for one furious man who stood up and demanded that the boys stop. When the doors opened at the next stop, the boys ran out. The victim stood up with blood slowly dripping from his nose. He had nothing but his hands to wipe it away with. Someone got up and handed

him a tissue. When questioned about how the fight started, the boy told his friend that they shoved him so he shoved them back. After he took a seat, there were still whispers about the situtation. It was soon clear that a gang chose to beat up this innocent person. I was still in shock. Why did people have to be so cruel? I couldn't help but stare at the helpless boy and feel sorry for him.

I wanted to know more about the gang problem so I paid a visit to the local police department and had a conversation with a New York City police officer, Mr. Smith, to get his perspective on gangs. Officer Smith is very familiar with gangs, including how they operate and how they are organized. In fact, he has seen three different people who were shot and killed. When asked why he thought people choose to be in a gang, Officer Smith replied, "I don't know. I think it gives them a sense of security and understanding." He thinks that most people involved are males from the ages of 13-25. "I think, if a teen is going to join a gang, it's going to be because they are pressured," said Officer Smith. "I feel sorry for them and wish they would try to give themselves something else to do."

How could a person treat another human being like that? When we walk down the streets, we sometimes glimpse things that could put us in shock. Some of the people responsible are gang members. Gangs are being formed everyday and are known for having specific attributes, like clothing styles and colors, hand signs, handshakes and certain areas that they mark as their territories. According to the National Youth Violence Prevention Resource Center, around the country there are more than 24,500 different youth gangs and 772,500 teens and young adults who are part of those gangs. They say that gang members are both female and male and are between the ages of 12 to 25. They also stated that male youths are more likely to join gangs than female youths, but a full eight percent of all gang members are female. These people are from different ethnic groups. Since the 1970s, there has been an increase in gang activities in the United States but, beginning in 1996, the number of gangs and gang members has decreased. It was shown that 20 percent of teens know someone who was killed due to gang violence. Gang members are also being murdered by other gangs. In some cities, 70 percent of teens that are murdered by guns are gang members.

If these teens and young adults are being killed because of competition among the gangs, why do they still continue to join these gangs? According to Cleveland Clinic, teens and young adults join gangs to have power, money, a substitute family, respect, security, protection, status, friendship and because they are bored. When a teen feels unloved at home, they may turn to other people who they think will show them the love that they long for. They turn to gangs because they feel like they will be protected and have a lot of friends.

The Cleveland Clinic also states that gangs often start out as a group of friends. While in these gangs, people make enemies and are

SUGGESTED RESOURCES

Books:

Scorpions by Walter Dean Myers

The Outsiders by S. E. Hinton

Autobiography of My Dead Brother by Walter Dean Myers

Always Running by Luis Rodriguez

The Girls in the Gang: A Report from New York City by Anne Campbell

Movies:

Westside Story

Colors (R)

Redemption: The Stan Tookie Williams Story (R)

Boyz In The Hood (R)

Dangerous Minds

Television:

The Wire, Season 1

Field Trip:

Rikers Island jail (New York City)

expected to be ready for violence and crime. Once you're in a gang, it's hard to get out. To turn your life around and walk away from gang life, you should find other things to do with your time, like playing a sport or attending after-school programs. When you make the decision to leave a gang, you shouldn't let the gang know because they might beat you or hurt you. Letting go of friends with negative influences, taking part in school activities, not using violence to solve your problems and being your own person will all help you avoid the gang life, according to the Cleveland Clinic.

I didn't know much about gangs—or why they are formed—before I researched this topic, but I have seen violence on the streets and have heard about gang violence. All I knew was that they are dangerous and everyone feared coming into contact with them. After doing research and interviewing Officer Smith, I learned more about gangs and why they are formed. "I think gangs are something that are uneccesary. There are a lot of things people could do other than joining gangs," he said. Gangs aren't necessary for someone to feel accepted and appreciated. The next time you think about being part of a gang, stop and ask yourself: Is it worth it?

Sources

"Health Extra- Gangs and Violence." The Cleveland Clinic Health Information Center. 2nd February 2007. http://www.clevelandclinic.org/health/

Jassiel Herrera

People might join gangs for protection and family.

"Violence is something people in the hood go through just to survive," said Jerrell, a gang member in East Harlem. Jerrell is my friend and he didn't have a problem telling the world the truth about being in a gang. He feels so comfortable and open with this issue because he feels people misjudge gang members. He wanted to give the perspective of a gang member himself. I never thought about myself being friends with a gang member because he is still the same person he was before he was in a gang.

Gangs and violence are something I have grown up with since I was young, but I've never been directly a part of it. Coming from the hood or projects makes it difficult to avoid gang activity or violence. That kind of stuff is a part of you whether you are in it, you witness it, or you are just at the wrong place at the wrong time. I, personally, never felt the need to join a gang because I already have a group of friends who support me and I wouldn't want to be one of those people who join something just to be cool.

I believe people join gangs for a lot of different reasons, but most of them are personal. Where I am, some people join gangs because of their cultural heritage. Another reason people join gangs is to get protection. There is a lot of violence in the hood and a gang can give you a perceived sense of security. I understand these reasons, even though I am not part of this, and that's why I'm willing to have friends who are in gangs. I have noticed that people who join gangs are often as normal as everyone else and I believe it's important not to judge them.

With this said, it is true that violence is a part of gang life. One day, I was walking down my aunt's block, close to where I live. I was wearing a blue shirt and hat. I ran into five members of a certain gang. They crowded me and asked me if I was a member of a rival gang. I told them no and I remember one of them saying I was lying. I wasn't scared because it wasn't the first time had I experienced something like that. I was ready to fight because I was raised in a violent neighborhood. Five minutes later, a friend of mine in the group recognized me. He told them who I was and they backed off. I felt like he saved my day because I couldn't take on all of them by myself. Growing up in an environment where there's a lot of gang violence, there are certain places where you just don't go by yourself. This was one of those places. In fact, there are many blocks where I can't go without the company of my friends. If you go by yourself, the kids there will try to beat you up. I wish it were different, and that I could walk anywhere with out watching my back all the time or worrying about my surroundings.

Jerrell has his own reasons for joining his gang. "Friendship, brotherhood, second family and protection," he states passionately. "Where I come from, the projects, it's not easy to stand by yourself." He also joined his gang primarily because another gang threatened to kill him. He had a friend who told him that he needed to join this gang so that the other one would not think he was on his own. Not everyone joins for such dramatic reasons. Jerrell said that some do it to impress girls or to be cool or because they are peer pressured. It's just another way to feel accepted.

While there are things a gang might appear to offer you, there are drawbacks. First, joining a gang can be difficult. "In some gangs, you have to fight numerous members from the gang before you're let in," said Jerrell. Also, violence is a part of their lives. Gang members are always getting hurt. And, according to the police, you're guilty by association. Whatever one gang member does, the whole gang is responsible.

Jerrell's perspective is interesting. He showed me that sometimes gang violence is done in defense and, as he pointed out, to save a life. "I still do things I used to do. I still attend high school and I also play basketball," he said, meaning that you can still have a life and do things you always do outside of your gang. I have great respect for him, but I'm thankful that I don't have the need or desire to join him.

Kevon Mitchell

They're a product of their environment.

Growing up in an urban lifestyle, young teens witness a lot of violence and gang activity. Depending upon where and how you grow up, violent lessons can be instilled in you and can become a part of your everyday life. While not everyone joins a gang, those who do have very specific reasons for doing so.

I interviewed Julio, a member of the Almighty Latin King And Queen Nation, who lives in Washington Heights. He explained his point of view on gang violence and why people join gangs. He even spoke about some of his past experiences with gangs. When I asked him why he chose to join a gang he replied, "I joined because I had beliefs in empowering my Latin people." Basically, he hopes to make a better way of life for his people. When I asked him why he thinks that people join gangs he said, "Some people feel that being in a gang is cool, some people are looking for a second family, and some people do it for the love and respect that they receive from their gang that they might not be getting from home."

Julio even went on to share his past experiences with gang violence. Once, he was walking through the projects in Harlem. It was two in the morning and he was on his way home, walking with a colored flag, when he was confronted by seven members of an opposing gang. He went on to say that he recognized one of the members. After they finished arguing, he continued walking home. When he reached his block, he noticed that they had followed him. He was once again confronted, but this time a fight broke out. While they were fighting, one of the opposing gang members pulled out a switch blade and swung at his face. He put his hands up to block and was sliced on the arm.

But that was only one of the many consequences he faced as a result of being part of a gang. He also stated that almost every weekend he was arrested. He looked at it as a sacrifice for his people. Meanwhile, his mother had given up on him and he was failing classes in school. He thought it was natural for him to sacrifice for a group that seemed to support him and make him feel successful. When I asked his girlfriend, Julia, how she felt about it, she said that it was the worst thing that ever happened to her. She would spend nights awake worrying about him and crying. She believes he is better as his own person rather than as a member of a gang.

Not all gang members are necessarily bad people. There are a lot of different people in gangs with different motives. Some think that it's cool and always want to fight. Those are the gang members that give gangs a certain reputation. Then there are the members with perceived hopes of raising up their people and helping those in need through gang association, like the large group of The United Blood Nation who protested alongside the Black Panthers following a police shooting in Queens. Then there are those who join gangs because they are scared and are in need of protection. For the others, they are often a product of their environment.

Questions:

1. What are the differences between a "gang" and other "groups"?

2. Why is being in a gang important to people? What makes teens join gangs?

3. Where does most gang violence occur? Why do you think this is?

4. What influence does rap music have on gangs, if any?

5. How should your school prevent gang violence?